Genesis of the Big Bang

GENESIS
OF THE
BIG BANG

Ralph A. Alpher

Robert Herman

OXFORD
UNIVERSITY PRESS

2001

OXFORD

UNIVERSITY PRESS

Oxford   New York
Athens   Auckland   Bangkok   Bogotá   Buenos Aires   Calcutta
Cape Town   Chennai   Dar es Salaam   Delhi   Florence   Hong Kong   Istanbul
Karachi   Kuala Lumpur   Madrid   Melbourne   Mexico City   Mumbai
Nairobi   Paris   São Paulo   Shanghai   Singapore   Taipei   Tokyo   Toronto   Warsaw

and associated companies in
Berlin   Ibadan

Copyright © 2001 by Oxford University Press, Inc.

Published by Oxford University Press, Inc.
198 Madison Avenue, New York, New York 10016

Library of Congress Cataloging-in-Publication Data
Alpher, Ralph.
Genesis of the big bang / by Ralph A. Alpher and Robert Herman.
p.   cm.
Includes bibliographical references and index.
ISBN 0-19-511182-6
1. Big bang theory.   I. Herman, Robert, 1914–   II. Title.
QB991.B54 A47   2000
523.1'8–dc21   00-023888

1 3 5 7 9 8 6 4 2

Printed in the United States of America
on acid-free paper

Preface

One of the reasons for writing a preface is to explain why the authors have felt moved to write a book in the first place. A second reason in this case is not only to arouse curiosity and interest in cosmology and its scholarship but also to stimulate potential readers to read *this* book, since there are others dealing with many of the same topics we cover.

Our book has had a long gestation period. We have had in mind writing about the development of the Big Bang model of the universe ever since a landmark event took place in 1965. It was in that year that Arno A. Penzias and Robert W. Wilson, then at the Bell Telephone Laboratories, published their discovery of a residual cosmic microwave background, the now-famous three-degree radiation, which pervades the universe. Almost overnight the scientific community as well as the lay public found reason to believe that a dynamic, evolving universe beginning with a singular event, the so-called Big Bang, was a credible model. This discovery was only the most recent significant step, because it had become clear many years earlier that the idea of an expanding universe was not only consistent with but virtually required by Einstein's general theory of relativity. Moreover, the observation of that expansion by the astronomers Vesto M. Slipher, Edwin Hubble, and Milton Humason had come to be widely accepted. Finally, it had also become evident that it was necessary to invoke an extremely hot, dense early stage in the expanding universe in order to understand the cosmic abundances of the lightest elements—namely, deuterium, helium, and lithium—relative to hydrogen.

In our view, the discussion of these concepts by Steven Weinberg in his book *The First Three Minutes*, first published in 1977 (the revised edition of 1988 is still in print), made a significant contribution to the widespread acceptance of the Big Bang model by many scientists, despite its targeting a general audience. Since 1977, other books have dealt with the model in more or less detail and with more recent observational and theoretical developments. Some of these books purport to convey the human side of the conception and development of the model as well. So, again, why another book?

We have been intimately connected with the conception and development of the Big Bang model since 1947. Following the late George Gamow's ideas in 1942 and more particularly in 1946 that the early universe was an appropriate site for the synthesis of the elements, we became deeply involved in the question of cosmic nucleosynthesis and particularly the synthesis of the light elements. In the course of this work, we developed a general relativistic model of the expanding universe with physics folded in, which led in a progressive, logical sequence to our prediction of the existence of a present cosmic background radiation some seventeen years before the observation of such radiation was reported by Penzias and Wilson. In addition, we carried out with James W. Follin, Jr. a detailed study of the physics of what was then considered to be the very early universe, starting a few seconds after the Big Bang, which still provides a methodology for studies of light-element nucleosynthesis.

Because of our involvement, we can bring a personal perspective to the subject. Our goal is to present a picture of what is now believed to be the state of knowledge about the evolution of the expanding universe and to delineate from our own unique vantage point the story of the development of the Big Bang model as we have seen and lived it. Our principal contributions have been our work on the synthesis of the chemical elements in the early stages of the Big Bang and our prediction of the existence and magnitude of a thermal cosmic background radiation with intensity peaking in the microwave region of the electromagnetic spectrum.

Because we were closely associated with the late George Gamow from the end of World War II until his death in 1968, we intersperse some reflections and anecdotal material about this remarkable physicist, his legacy in science, and the influence of his personality and image on the way our joint and individual contributions in cosmology have been perceived by the scientific community. Gamow was a prolific writer of books interpreting science for the layman, which conveyed to his readers his enjoyment in doing science. Because he wrote on physics and cosmology at a popular level, and because he injected a considerable amount of humor into his presentations, he was frequently not taken seriously by too many of his fellow scientists. His not being taken seriously is something that rubbed off on the two of us as his colleagues, particularly because we were working in such a speculative area as cosmology. Nevertheless, a number of current colleagues attribute their early stimulation toward a career in science to his popular books on physics and cosmology, particularly *One, Two, Three . . . Infinity* (1947).

Our book is intended for a rather broad audience, although we do not pretend to have Gamow's skills in popular writing. We attempt to deal with technical details in the text through descriptive explanations, to the best of our ability; some of the relevant mathematical material is treated in the appendix.

Since we want to emphasize primarily that segment of the subject in which we were personally involved, we have not described in any depth the increasing amount of recent research that attempts to combine high-energy particle physics with cosmology at extraordinarily short times after the singular event called the Big Bang. This aspect of the field is in a much greater state of flux than the development of the standard model on which we focus, although there have been successes in dealing with the physics of these very early times or at least in understanding some of the questions previously left unanswered by the standard model. Since we have not been personally involved, however, we leave it to others to write historical accounts of recent concepts, particularly as the connection between theory and observation continues to improve. Moreover, we address the exciting and voluminous research being done under the rubric of observational cosmology only when it is particularly cogent to the various subjects we discuss.

The organization of the book is as follows. First we present some historical background to develop a context and then go on to a coherent account of what is known as the standard Big Bang model, so that the reader will have an idea from the start of what Big Bang cosmology is all about. In subsequent chapters we address the conception and development of the model in what we hope is an orderly manner, although it is our experience that scientific developments appear to be orderly only in publications that review a field of activity. We deal very briefly with some of the proposed alternatives to the Big Bang model. On occasion we inject a remark or two on our perception of how science appears to be done, based on the material and story we have by then conveyed. A selected reading list of some current books and a few relevant articles on cosmology is appended.

This book was completed by Alpher, since Robert Herman, his collaborator for over 50 years, died on February 13, 1997. We had fortunately worked through an outline of the substantive content of the book and had even developed first rough drafts of a few chapters. By and large, however, what appears here is the work of Alpher, with a flavor that he hopes would meet with Herman's approval.

Finally, we reiterate that this book is not intended as a treatise on modern cosmology but rather as a window on the experiences of two collaborators in the critical areas of primordial nucleosynthesis and the cosmic microwave background radiation.

RAA

RH

I owe much to many: to the late George Gamow for inspiration in physics and cosmology, and for a lot of perspiration; to the late Robert Herman, who, after 50 years of collaboration and shared perspiration and frustration, passed away as this book was beginning to take shape; to my wife, Louise, son Victor, daughter Harriet, and son-in-law Glen, and to Helen Herman, for their continuing support as Herman and I suffered through many years of the ups and downs of our involvement in Big Bang cosmology (Victor was kind enough to read the book in manuscript and supply insightful comments). James Rosenthal provided computer skills in the preparation of the mathematical appendix and illustrations. George Wise helped on some historical questions during the final stages of preparation. I am pleased to acknowledge help in the final editing of the manuscript by Philip Koske and Donald R. White. I am indebted to the Physics Department of Union College for the home, facilities, and sympathetic assistance it has provided over the last decade, and to several people from its Office of Computer Services, who assisted in my problems with dealing with a computer that I frequently overloaded and otherwise misused. Finally, the book has been much improved by the suggestions of Kirk Jensen, Executive Editor, and the production staff at Oxford University Press.

RAA

Contents

Genesis of the Big Bang

Herman (*left*), Gamow (*center*), and Alpher (*right*). Alpher took the photo of
Herman holding a wired programming plugboard for an IBM CPC computer at
IBM's Watson Laboratory (then near Columbia University). Atomic physicist
C. L. H. Thomas, a resident scientist, assisted with programming. The photo of
Alpher was taken by *Newsweek* but never used. Gamow's photo came from a
security badge at the Applied Physics Laboratory, Johns Hopkins University.
The bottle was labeled YLEM by Gamow (the word, found by Alpher in a large
dictionary, means the primordial stuff from which all matter was formed), and
its contents were partially consumed by those in the photo to celebrate the
mailing of the Alpher-Bethe-Gamow paper to *Physical Review* in 1948. Alpher
and Herman slipped the montage as a slide into a collection; Gamow first saw
it while delivering a lecture in 1949 at Los Alamos.

1

An
Overview of the
Big Bang Model

Genesis of the Big Bang examines the development of the standard Big
Bang model of the universe. We were personally involved in the early
stages of this development and believe that we can supply some unique
insights.

In this first chapter we hope to make it easier for the less technically
oriented reader to follow our subsequent discussions by providing an
overview of what the Big Bang model is about. For those with more
sophistication in mathematics and physics, there is a brief mathematical
discourse on various topics in the standard model in the appendix. By
no means is the appendix intended to be a treatise on cosmological
models—but some readers will, we hope, find it useful. In the appendix
mathematical results are generally stated rather than derived, since there
are a number of treatises on cosmology containing detailed derivations
of the relevant material.

Our overview is presented without any particular attempt to place the
subject in a historical or sociological context. We remark on the nature
and credibility of the underlying assumptions and the supporting evidence for the Big Bang model. Our description of the very early universe
(starting at much less than a microsecond after the Big Bang) is sparse
but will, we hope, be informative. Excellent descriptions of this aspect of
the subject are already available in other sources. Moreover, in contributing to studies of what is now considered the moderately early universe
in order to clarify the initial conditions for the synthesis of the chemical

elements early in the universal expansion, we attempted to keep the necessary assumptions as simple as was consistent with the development of a useful and successful theoretical model. Generally, it is only as a last resort that scientists invent new phenomena or introduce additional assumptions to bring a model description into closer agreement with observation. We make no claim that a description of the universe should necessarily be simple or beautiful, but it would certainly be more aesthetically and intellectually satisfying if it were so; the goal of much scientific work in many disciplines is parsimony in modeling. According to Gamow, the famous physicist Paul A. M. Dirac is said to have remarked that if a theory was beautiful, it was probably right.

Physical cosmology confines its attention to the "how" of the universe and does not deal with the "why." The universe is taken to be the observable ensemble of elementary particles, atoms, ions, molecules, planetary bodies, electromagnetic radiation over all wavelengths (from gamma rays and x-rays through the visible light spectrum to the infrared and on into radio frequencies), neutrinos (massless or, as now seems more likely, very low-mass exotic elementary particles that are important in astrophysics and nuclear physics and may be quite important in cosmology), gas, dust, stars, clusters of stars, galaxies of stars, and clusters of galaxies—and clusters of clusters of galaxies, as well, if there is a next step in hierarchical structure. There is also a mysterious component, in an amount yet to be determined, called "dark matter." It is an awesome ensemble.

Much of the gas is in a neutral plasma state. It consists of positively charged ions—that is, atomic nuclei stripped of some and perhaps all of their electrons (depending on the ambient temperature)—and free negatively charged electrons, with the sum over all positive and negative charges in any significant volume of space being zero, according to an assumed principle of conservation of charge, which appears to pertain. In other words, the cosmos is on the whole electrically neutral. The composition of the universe is about 75% hydrogen, 24% helium, with very small or trace amounts of all other species of atomic nuclei.

Cosmologists assume that the fraction of the universe we see is representative of the whole. This assumption was first made seriously at the time of Einstein, and it has seemed to hold up as observers become able to penetrate more and more deeply into the observable universe. According to essentially all proposed models, astronomers now observe only a fraction of what we shall ever be able to observe. We do not know how large this fraction is. The associated upper limit to the distance is defined by a horizon which is at a distance equal to $c \times t_o$, where c is the velocity of light and t_o is the elapsed time since the Big Bang event. There are current observations of objects whose distance is estimated to be more than about 0.8 of the radius of the potentially visible universe—that is, that portion of the universe which is accessible to us according to the laws of nature. Since the volume of a sphere varies as the cube of the radius, this means we are seeing about 0.5 of the volume of the potentially visible universe. If somehow we determine that this fraction is not globally representative, then the task of modeling the universe will be much more difficult. If the universe went through a period of inflation commencing with the Big Bang event, it is part of the inflationary paradigm that the universe at all times exceeds in size what is observable. This makes it reasonable to suppose that the universe as seen on a relatively small scale appears to be flat. (We shall say more about flatness in chapter 6 on the inflationary paradigm.)

The assumption of a fair sample is not a trivial conjecture, for astronomers have been finding such large-scale structural features in the universe as to stretch the validity of this assumption. Now on scales exceeding about 300 million light years, the universe appears to be smooth. Recall that a light year is the distance that light travels in vacuum in a year, at a velocity of 3×10^{10} centimeters per second, namely 9.46×10^{17} centimeters, or about 63,000 times the distance of Earth from the Sun. If the universe is 15 billion years old, then the distance to the observable horizon is 15 billion light years, and 300 million light years is a small fraction of that.

Perhaps the most important and essential assumption made in cosmology is that the universe is understandable. Some cosmologists suggest that its most interesting aspect is that it exists at all. Even if one finds the universe indifferent to human existence, chaotic, and operating in a probabilistic manner, it is nevertheless thought not to be "capricious," not playing games with us—a point strongly emphasized by Albert Einstein and widely quoted. Yet nowhere is there a guarantee that a *full* understanding of the universe is either simple or even possible, now or in the future. There is an interesting quotation usually attributed to J. B. S. Haldane, a British scientist: "My suspicion is that the universe is not only queerer than we suppose, but queerer than we can suppose."

Einstein's well-tested theory of special relativity provides cosmologists with the ability to effectively travel back in time. The theory of special relativity postulates that the velocity of light is both finite and constant, namely, 3×10^{10} centimeters per second, as already mentioned (equal to 186,000 miles per second). Moreover, a very surprising aspect of the travel of light is that its velocity is independent of the velocity both of the emitting light source and also of the observer. As far as we know, the most rapid possible way to transmit information is at the velocity of light. For example, the Sun is at a distance from Earth of about 93 million miles. If the Sun were to be snuffed out suddenly, we would not know it until light no longer reached us, and this would take about eight minutes (dividing the distance of the Sun, 93,000,000 miles, by 186,000 miles per second the velocity of light, gives about 500 seconds or 8.3 minutes). It takes about four years for light to reach us from the star nearest the Sun. If the universe is assumed to be about 15 billion years old and was somehow producing light from a time near the Big Bang, then we could in principle observe at most only to a distance of 15 billion light years (i.e., 15 billion years $\times c$, where c is the velocity of light). The volume defined by this distance in turn defines the potentially visible universe. (We will have a bit more to say about the age of the universe in chapter 3 and the appendix.)

So it appears to be that the observable horizon grows with the aging universe at the velocity of light. Thus, when the universe is a billion years older than it is now, it should in principle be possible to see objects one billion light years farther away—that is, at a distance of 16 billion light years. To emphasize the significance of our remark about time travel, let it be noted that if we were to observe a celestial object at a distance of, say, 10 billion light years, then we would see it by light emitted 10 billion years ago, when the object was much younger, less evolved, and perhaps quite different in appearance and closer to its neighbors, if the number of such objects in a given volume was higher in the past. This is expected behavior in an evolving universe. The actual determination of age from distance measured to an object is model-dependent; except for nearby objects, the imputed age will depend on whether the universe is open, closed, or flat as will be clear from the brief discussion about the several models elsewhere in the book and the calculation of cosmological age in chapter 3 and the appendix.

The basis of the Big Bang model is the assumption that the universe was hotter and denser in the past, and that the universe and all the objects in it have formed and evolved from that earlier state. Thus, it is an important aspect of astronomical research to try to understand how the light output of galaxies changes as stars are born, live, and die within the galaxy or galactic cluster whose collective light output we study. All stars have a finite lifetime, and their characteristics change as they age. To reiterate, the theory of special relativity provides a kind of time travel into the past. When we look out into space, we look back in time. This is both an opportunity and a limitation.

Observations coupled with the theory of special relativity indicate that the universe is *homogeneous* (the same everywhere) and isotropic (its properties are the same in all directions) on the grand scale, which leads to alternatives for its geometry. The geometry of the universe consistent with these characteristics is one that is either infinite in extent or else finite but unbounded. One way to think about an infinite extent is to realize that if it were finite, then one would have to contend with the idea that there is something beyond—an idea that contradicts the concept of *universe* we are pursuing, where we take the the word to mean "all there is," including parts that we are not yet able to observe. The possibility of the universe as finite but unbounded can perhaps be visualized by imagining that all the objects in the universe are flattened and pasted to the surface of a sphere. A resident of this universe could move in any direction on the surface without limit, for there would be no edges. Such a universe would be clearly finite in size but with no boundaries, and residents would have to do clever experiments to deduce that they were living on a spherical surface—including deducing the location of the center of its curvature, which is technically outside the space-time domain of the observer on the surface.

There are statements in the early literature on relativity theory that if an observer in a closed universe, finite but unbounded, were to remain at the same spatial coordinate, then he could see the back of his head after a time delay because of the velocity of light. Such boomeranging of the light will not in fact occur because, during the passage of the light, time runs on everywhere and at the site of the observer. Something like this could occur, however, in the intense gravitational field near the strange objects known as *black holes*. A black hole is an evolved star that has collapsed to such a small radius that its surface gravity is very great, so great that light cannot escape and climb out of the gravitational field. Black holes in principle may be indirectly observable through their gravitational effects. For example, one might search for very high velocities among stars trapped in and rotating about presumed black holes in what are called accretion disks. Moreover, evidence is accumulating that we are seeing high-energy radiation from material in accretion disks, generated as the material swirls around the parent body and is accelerated toward it. Another effect might be the external manifestation of strong magnetic fields, should they exist, and they almost certainly do. Essentially all stars have a magnetic field, and when one collapses into a black hole, its magnetic field is frozen in to the collapsing material and is concentrated; thus, the resultant black hole should have a very high surface magnetic field. The material of the black hole is nearly a perfect conductor during the collapse phase, and magnetic field lines will be locked within such a conductor. The collapse then leads to a concentration of the magnetic field lines. If the black hole has a net positive or negative charge, there would be some external effect. One more effect, predicted by theory, has to do with the possibility that black holes once formed do not last forever but lose mass. This is carried off as energy from the strong gravitational field environment very near a black hole by the momentary creation of electron-positron pairs: one particle escapes to infinity (i.e., propagates away), while the other is reabsorbed in the black hole; the result is a net energy loss whose rate is dependent on the mass of the black hole. This phenomenon was first suggested by Stephen Hawking some years ago but has yet to be verified observationally.

In the language of relativity, we are concerned with events in space-time rather than coordinate points specifying location in space alone. One can locate the position of an object by specifying three numbers associated with the usual three spatial dimensions, but an *event* associated with this object requires specification of a fourth number: the time elapsed since some fiducial time. This fourth number can be considered an additional dimension by which an event is defined. If the universe is thought of as a four-dimensional volume (with three dimensions of space and one of time), then it could in principle be viewed in toto from a still higher dimensionality—namely, in five dimensions. In general relativity and in the Big Bang model, however, we speak of a four-dimensional space-time as the defining geometry. As we explain in chapter 3, this space-time has additional geometrical properties that depend on the distribution of matter and energy in the universe and its rate of expansion.

Cosmologists refer to the idea of the homogeneity and isotropy of space as the "cosmological principle." If, in addition, the cosmos is unchanging with time, a concept on which the now out-of-favor model of a Steady State Universe is based, one refers to a broader "perfect cosmological principle." As of this writing, observational cosmology finds nothing inconsistent in the assumption of homogeneity and isotropy on a scale of, say, 300 million light years.

Homogeneity and isotropy on a large scale lead to the statement that all observers are equivalent, wherever and whenever they may be. There are no special or unique observers or places of observation (there is a possible exception in the sense that to each observer in the cosmos there corresponds a *surface of last scattering*, which has to do with the locale or origin of the free-running cosmic microwave background radiation, discussed further in chapter 5). To understand what is meant by a large or cosmological scale, we must expand our view and talk about galaxies, or even clusters of galaxies, as the fundamental building blocks of the cosmos. The notion of clusters is an important one. Most galaxies, including our own Milky Way, are found in nature to be grouped together in numbers of 1,000 or more, held together by mutual gravitational forces. Consider that the universe is filled with galaxies, separated from one another by distances of the order of several millions of light years, and by clusters of galaxies, occupying even larger volumes of space. These galaxies are like our Milky Way in that each contains billions or tens of billions of stars.

The Milky Way is an object whose visible constituents are stars, gas, and dust, as well as clusters of stars distributed in a disk through a volume of about 10,000 light years in thickness and 100,000 light years in diameter. This huge assembly rotates about an axis perpendicular to the plane of the disk once every quarter of a billion years. The disk's edges are slightly warped, rather like the brim of a hat. It may also contain a large amount of so-called *dark matter*, in a more or less spherical halo, whose existence is inferred by astronomers primarily by its gravitational influence. Dark matter has not yet been observed directly by any electromagnetic radiation it might produce. However, recent work—not yet verified—suggests that a large fraction of the missing mass, or dark matter, is fully ionized hydrogen in intergalactic space, inferred from the population of highly ionized oxygen atoms. The hydrogen absorbs indirect energy from galactic clusters and emits radiation which is not visible but which produces visible emissions from ionized oxygen. The Sun and its system of planets occupy an undistinguished position in the plane of the disk, about two-thirds of the way out from the axis of rotation. Were one to view the Milky Way from a considerable distance, one would find the various luminous constituents concentrated into arms, giving the galaxy the appearance of a multiarmed *spiral*. There is, moreover, a short linear feature in the distribution of stars at the core of the galaxy which identifies the Milky Way as a *barred spiral*. Astronomers point out that except for this recently discovered bar the Milky Way should look much like the Andromeda Galaxy (very familiar galaxy to aficionados of astronomical photography), which is at a distance of some two million light years.

Another major type of galaxy, called *elliptical*, is also present in considerable number in the cosmos. These are comparable in size to spiral galaxies but do not exhibit the arm structure of a spiral and contain little gas or dust. There is also a class of galaxies called irregulars. Whether spiral, elliptical, or irregular in form, galaxies seldom occur alone but, rather, are found in clusters with as few as several dozen and as many as a thousand or more members. Such clusters appear to be the basic building block of the universe. The Milky Way is part of a local group of more than 20 galaxies, some of which were first identified only in the last few years.

The foregoing list by no means exhausts the types of galaxies. Perhaps most fascinating are recent discoveries of interacting galaxies, apparently seen in collision. Since we do not want to compete with the many books and, in particular, the popular astronomical magazines that do a superb job of color reproduction of cosmic photographs, we go ahead with other matters. Finally, we reiterate that when we speak of large scales on the cosmological level, we are speaking of distances that encompass a number of clusters of galaxies, at least several hundred million light years.

The concept of an average density of matter in the universe was introduced by Einstein in his first foray in cosmological modeling. By this we mean that if the total mass of matter plus the mass equivalent of the radiant energy in any given volume of space were smeared out uniformly through the volume, then the average density is that total mass divided by the volume. In other words, the average density of matter plus energy is what would obtain if all the material collected in stars, dust, and gas in the clusters of galaxies, together with the radiant energy and unseen mass in the volume, were somehow redistributed as a uniform cloud throughout the relatively finite space occupied by the clusters.

The laws of nature are assumed to be valid everywhere. They are scientific constructs that describe the behavior of matter and energy. The laws that have been identified and studied by scientists in their laboratories or within the local solar system appear to be equally valid elsewhere and everywhere and at all times in the universe; that is, they are valid at all scales of observation, from the basic constituents of matter to the majestic clusters of galaxies, from nearby in space and time to great distances and times in the past representing significant fractions of the age of the universe. This assumption of the universal validity of the laws of nature can in many instances be tested by observation. For example, the laws governing the emission or absorption of light by nuclear species or by molecules in objects five billion light years distant have been compared with those governing the emission or absorption of light by the same species in the laboratory. The governing laws, which are unique and characteristic for every element, are remarkably the same. Despite overwhelming evidence for the validity of the laws of nature everywhere, however, it is consistent with the scientific process, and appropriate, that such statements be reexamined as our observational capabilities improve with time. In early 1999, for example, observations were reported, but have not yet been verified that a combination of universal constants known as the *fine structure constant* may be slightly different at large distances of the order of, say, five billion light years. At present, these observations set a very small limit on the departure of this constant from its terrestrial value, but should they be verified, there could be profound effects on cosmological modeling. Recently Kenneth Brecher (*Bulletin of the American Physical Society 45* [2] [2000] #34) presented evidence based on signals from a gamma-ray burster (origin of bursts still a mystery) that tests the second postulate of special relativity to $1:10^9$ better than previous tests, already at $1:10^{13}$. This is the postulate that states that velocity is independent of the relative motion of source and observer. This really establishes the velocity of light as a fundamental constant.

In the final analysis, all that the scientific process can provide in understanding any physical phenomenon is presumably a construct of the human mind: a theory or a model which, when coupled with observational data, makes the theory credible. The Big Bang model is based on Einstein's theories of special and general relativity and incorporates our understanding today of the behavior and evolution of matter and energy. (As we mentioned earlier, before the phrase "Big Bang" came into widespread use, we characterized the universe in our publications as "dynamic and evolving.")

The theory of general relativity models gravitation by replacing it with geometry. Gravitation is the major long-range force that controls the structure of the universe on astronomical scales, as well as locally preventing us from being flung into space from the surface of the Earth. In addition, our understanding of nuclear physics and the physics of the elementary constituents of matter provides a basis for understanding the composition of the universe and the details of its evolution.

The assumptions we have described appear to remove from consideration the possibility of a central position for *Homo sapiens* in the cosmos. This means that any observer, wherever located in the universe, will see the same panorama when viewing the cosmos on a large scale. Consider how little of the universe could possibly be in contact with us or have received information from us: that is, how small a volume of space has been traversed by light carrying information about the existence of *Homo sapiens*. Given that light is the information carrier, knowledge of the ancient Egyptian pharaohs is confined to a sphere of radius 5,000 light years, which is less than 1% of the volume of the Milky Way itself; manmade radio and television signals occupy a sphere of 70 light years' radius, which contains perhaps a few hundred stars. By and large, signals carrying the programs of Bob Hope, Milton Berle, Ed Sullivan, Dan Rather, Mike Wallace, the war in Vietnam, and other electromagnetic signals from our early forays into television broadcasting have not yet reached very far out on the cosmic scale, though they may some day be picked up by another civilization in the cosmos.

We argue for the Big Bang model as the most persuasive and inclusive physical theory of the cosmos at this time because the model has predictive ability (i.e., it encompasses simultaneously many and diverse astronomical observations) and in particular because, as any viable theory must, it continues to survive the challenges of observational falsification. The model may turn out to be wrong or wanting as the result of predictions or observations yet to be made; one must accept that science frequently progresses when existing ideas are challenged by logic or data and must be modified or altered by new concepts. To put it simply, a basic attribute of good science is that it is dynamic, evolving, and—as argued by the philosopher of science Karl Popper in his book *The Logic of Scientific Discovery* (New York: Basic Books, 1959)—self-correcting. Scientific research involves a theory or model of what is being studied and the interaction of that theory with experiment, which in astronomy is almost entirely observation. Observational cosmology does not lend itself to laboratory control, or to measurements under controlled conditions, but, to be credible, observations must be repeatable, and repeatable by independent observations. This is an established standard. A model is not reality but, to reiterate, a human, intellectual construction.

For example, the Darwinian theory of evolution of living organisms provides a framework for study, organizing observations, and prediction which has survived for a long time. Since the theory is a scientific artifact, it is surely not productive to criticize evolution in a highly pejorative sense as just a theory. Theories and models, which change and evolve, are, in the final analysis, all we have to help us organize our observations and explain the world in which we live.

In the case of the Big Bang, not only has the model survived over a number of decades, but the case for it has become progressively stronger. The most important defining observation is the well-known expansion of the universe, discovered some 70 years ago. Another defining event in the acceptance of the model by the scientific community was the 1965 confirmation by Arno Penzias and Robert Wilson of the cosmic microwave background radiation, whose existence had been predicted. Yet another defining event is the success of theoretical prediction of the relative abundance of nuclear species in the framework of the early Big Bang model. In summary, at the present time the Big Bang model appears to be alive and well, but it must be remembered that theories and models are always provisional and represent ongoing developments in scientific understanding.

The major difference between cosmology in the twentieth century and the earlier views is the idea that the cosmos is dynamic and has evolved from a very hot, dense early state. It is the prevailing view among cosmologists today that objective data do not support the argument that the cosmos is static and unchanging from "everlasting to everlasting."

Evidence for the Model

Indirect Evidence

Although the physical and astronomical evidence for the current Big Bang model of the universe is convincing to most cosmologists, some pieces of evidence may be less direct than others. Indirect but significant evidence is found in the observation that on a large scale, many millions of light years in extent, the universe appears to be homogeneous and isotropic, concepts that we have already mentioned. This, together with the universality of the nuclear composition of the universe on a large scale, suggests a common origin for its major features. It is important to realize that after Einstein introduced the theory of general relativity, it was found to incorporate nonstatic models of the universe. This alteration in Einstein's original assertions was developed in the 1920s by the mathematician Alexander Friedmann in what was then the USSR, by the cosmologist and cleric Georges Lemaître in Belgium, and by the astronomer Willem de Sitter in the Netherlands. Their work, which has since provided the basis for modeling the large-scale structure of the universe, argues against a static or unchanging universe and for a dynamic or evolving universe, one that expands, or expands and then contracts, or goes through a series of expansions and contractions (i.e., oscillations)—all characterized as evolutionary.

Other phenomena also suggest an evolving universe. Modern observational techniques enable one to estimate the age of the chemical elements from the present existence and relative amounts of radioactive nuclear species. The radioactive species uranium, thorium, and others exist in nature and undergo spontaneous decay (nuclear transformation) at a precisely measurable rate, from one type of nucleus to another, usually through a sequence of steps culminating in a nucleus that is stable against further change. Knowing the rates of radioactive decay, and the relative quantities of radioactive and stable isotopes, one can estimate when the species were originally formed. Isotopes are nuclei that have the same chemical properties but differ one from another in their mass because of differences in their neutron content. For example, uranium 238 and uranium 235 are isotopes, both radioactive and each containing 92 protons but differing principally in that there are three fewer neutrons in uranium 235. That radioactive species exist at all suggests that either they were formed at a finite time in the past or else they are being created continuously in existing structures such as stars; both possibilities argue for an evolving universe.

The evolution of stars is also an indicator of a finite cosmic age. Stars are formed from the condensation of gas and dust under the action of mutual gravitational forces. When embryonic stars become sufficiently condensed, the energy and pressure at their cores cause nuclei to react with one another, the energy thus generated escaping from the star in the form of electromagnetic radiation and neutrinos. The energy resulting from the nuclear reactions in the stellar core also maintains the star in a state of balance so long as there is nuclear fuel to supply energy, for the gravitational forces would otherwise tend to continue the collapse. There are good theoretical models of stellar structure and dynamics which provide an estimate of how long stars of a given type can survive before exhausting their nuclear fuel. These models indicate that some massive, bright stars live for only millions of years, whereas smaller, less luminous stars such as our Sun may live quite uneventfully for billions of years. This view represents a major change in our understanding of energy sources in stars before the introduction of nuclear phenomena. For example, in early studies known now to be wrong, the famous British astronomer Arthur S. Eddington's estimate of stellar lifetimes at some millions of years was based only on the release of gravitational energy as stars slowly contracted. Studies of the properties of stars give us a view of the age distribution of stars, much as studies of a human population can provide a view of the average age and the distribution of the ages of human beings around the mean.

In addition, stellar ages can be derived for the members of several types of stellar clusters in the cosmos. Clusters of stars within galaxies, as well as clusters of galaxies, are bound together by their mutual gravitational attraction. Given the proclivity of such collections to disperse (because they are moving with respect to one another, and because rapidly moving members will—by energy transfer from other members of the group— tend to achieve a velocity greater than the escape velocity, making it energetically possible for the fast members to leave the collection despite gravitational attraction of the total mass of the cluster), it is possible to calculate how long such clusters have been capable of surviving as recognizable entities. The times are consistent with a finite age of the cosmos, by no means forever. Parenthetically, such calculations suggest that there is more mass in galactic clusters than meets the eye as luminous matter, as first proposed by Fritz Zwicky in the mid-1930s. This is also the case within galaxies, particularly spirals, as shown by Vera Rubin and Kent Ford in the 1970s. Astronomers are seeking to map the distribution of this nonluminous matter in galaxies and clusters by observing the deflection of light from background objects, which is due to gravitational effects of all intervening matter, luminous or nonluminous.

Somewhat less obvious but nevertheless convincing is the evidence derived from astronomical radio sources and quasars. For the many sources of radio emission, which are undoubtedly galaxies, if one uses a radio telescope to count the number of such objects that are radiating at given intensities, one finds in a given volume of space there are more of such objects at greater distances. Given the finite velocity of light, we are seeing these greater numbers of objects in the past, suggesting that the separation of radio sources was smaller in the past and, therefore, that the universe has evolved from a denser state when there was a higher number density of such objects.

The class of objects called *quasars* are thought to be very luminous galaxies probably housing black holes at their centers, seen at great distances while they were quite young. Radiation characteristic of the elements hydrogen and helium can be seen as emitted or absorbed by these objects, but the radiation is greatly reddened—that is, wavelengths are shifted toward longer or redder appearance, compared with similar emission or absorption of hydrogen or helium in the laboratory—and this reddening signifies that these objects are receding. There is no convincing physical explanation for the reddening or *Doppler* effect (an apparent motion of the light-emitting source away from the observer) other than the expansion of space itself. A *redshift*, in simple terms, denotes the fact that since the velocity of light is a constant regardless of the motion of source or observer, a relative increase of distance decreases the frequency and increases the wavelength of the light. The basic formula is

$$\text{light velocity} = \text{constant} = \text{frequency} \times \text{wavelength}.$$

The relatively large amount of reddening indicates that quasars are among the most distant objects we observe. As further corroboration of their distance, there is now evidence of the phenomenon known as *gravitational lensing*, which was predicted by general relativity. The first observational test of general relativity was the deflection of light from stars seen close to the limit of the Sun during a solar eclipse—"gravitational lensing" by the Sun. This feat was accomplished by an expedition led by astronomer Arthur Eddington, in 1919. According to the physics of this phenomenon, the light from distant objects is deflected and focused by the gravitational fields of intervening objects such as stars (as by the Sun), gas clouds, and galaxies, producing interesting observable images and patterns that provide information about the distance of the light sources and the nature of the lensing object. Remember that what is actually interpreted as a deflection is the trajectory followed by photons of light in the curved space-time associated with the gravitating bodies. The existence and numbers of the quasar images, as with the radio sources, are consistent with the Big Bang model.

We note that gravitational imaging may contribute to our knowledge of distance scales in the universe, and we will say more about this later in chapter 3. Recent observations suggest a phenomenon called *micro-lensing*, which makes it possible to study nearer objects: in our galaxy, by observing local gravitational effects on the light coming from stars in a nearby galaxy such as the Large Magellanic Cloud. This effect is evidenced by a momentary flickering, a brightening or shift in the apparent position of a distant star as it passes through the curved space-time associated with a star in our own galaxy. Such observations should lead to better estimates of the quantity and nature of nonradiating objects that contribute to the dark matter content of astronomical objects.

More Direct Evidence

At least three kinds of observations are regarded as strong evidence for the Big Bang model. First and most important are the famous observations by the astronomers Vesto M. Slipher, Edwin P. Hubble, and Milton S. Humason at Mount Wilson Observatory in the 1920s and 1930s. They found that on a large scale, galaxies are observed to be receding from one another. Observers everywhere in the universe will see the same expansion effect. The word *everywhere* is significant, for it distinguishes the Big Bang from the simple and intuitive but irrelevant model of an explosion from a finite source. The correct physical picture is that galaxies and gravitationally bound clusters of galaxies are embedded in space-time, which is itself expanding. Galaxies that are located in this matrix and are now farther away from any observer had to have been traveling at a higher speed away from the observer to arrive where they are now. Thus, there results the famous velocity-distance relation known as the *Hubble law*, in which the apparent recession velocity of galaxies is proportional to their distance from the observer. The effect of expansion within, say, a local system such as the solar system, or even a galaxy or a cluster of galaxies, is overwhelmed by the larger gravitational attraction in local systems.

In the more than 70 years since this result was first published, no acceptable alternative physical explanation has been developed for the reddening of the light. The shift toward longer wavelengths on which the Hubble law is based can be characterized as a Doppler shift arising from the expansion of space, and appearing to the observer as a motion of the emitter away from the observer. Again, we emphasize that in terms of general relativity, what is actually expanding is space, or the classical measuring rod that one uses to measure distances. A simple exploding bomb analogy is inapplicable because it is inherent in the model that although the cosmic expansion had an origin in time (it began at and indeed acted to define the zero of time), there was no preferred location in space for an infinite or unbounded universe, consistent with there being no preferred observers. Hence, the expansion occurred everywhere at a given time.

Cosmologists frequently try to clarify the nature of the expansion by using a raisin bread analogy. Consider a mound of yeast dough in which raisins are randomly dispersed. The dough rises as the yeast generates bubbles of carbon dioxide. Imagine yourself on one of the raisins. You will observe the other raisins moving away from you at velocities proportional to their distance from you. So also would an observer on any other raisin. If the crust is sufficiently far away, all raisin-resident observers will be equivalent, and no unique center from which the bread rises can be identified by an observer on any raisin.

To use another common analogy, visualize the universe as the surface of a balloon on which galaxies are distributed as spots. As the balloon is inflated, the spots will move away from one another. Clearly, there is no preferred position on the surface, and the spots more distant from an arbitrarily selected position on the surface of the balloon will have to move faster as the balloon is inflated in order to maintain the balloon geometry. Note that the center of the expansion (i.e., the center of the balloon) is not in the space-time geometry of the two-dimensional residents of the two-dimensional surface of the balloon. With suitable local measurements of curvature, however, the existence and location of the center can be inferred by surface-bound observers.

In considering the age of the universe, by which we mean the elapsed time since the Big Bang event, it is at least provocative that there are now two schools of observational cosmologists. One school claims that the evidence points to an age of the order of 10–12 billion years—which is less than the derived ages of some objects in the universe. This is a clear contradiction; if this school is right, then the standard Big Bang model will have to be modified. The other school finds the expansion rate determined from available techniques to give an age of 15 to 20 billion years, consistent with the age one can calculate from the Big Bang theory and more than the age of any structures in the cosmos. In the Big Bang model, this calculation requires a knowledge of two parameters: one, the rate of expansion of the measuring rods of the universe, given by the *Hubble parameter*, which is the constant of proportionality between recession velocity and distance; the other, the average density of matter in the universe. (Mathematical expressions for calculating the age of the cosmos in terms of these parameters, and some graphical illustrations as well, appear in the appendix.)

A simple calculation based on the incorrect bomb analogy requires one to assume that the flight or recession of galaxies from one another is not unlike the dispersion of fragments from an exploding bomb. If there are no retarding forces on the fragments—such as the air drag that affects an explosion in the earth's atmosphere—then the velocities are unchanged from the time of the explosion. From measurements of the velocities, one can calculate back to determine the time of the explosion. This result, applied to the cosmic expansion, yields the so-called *Hubble age*, an approximate determination of the age of the cosmos that is intuitively simple but not precise. There is current uncertainty not only about the Hubble parameter but also about the mass density if all matter and radiation were smeared out uniformly through space. The density determination is complicated by the presence of a significant amount of matter inferred only from gravitational effects but not yet observed directly. There is an indirect measure for dark matter involving gravitational lensing, as already mentioned. To calculate the model age more precisely requires the solution of the relativistic equations underlying the model.

The second major piece of evidence has to do with the observed relative abundances in the universe of the light chemical elements hydrogen (^{1}H) and its isotope deuterium (^{2}H), helium (^{4}He) and its isotope (^{3}He), and, according to very recent studies, the element lithium (^{7}Li) and its isotope (^{6}Li). It now appears that these elements, perhaps along with small amounts of beryllium and boron (collectively called the light elements), must have been synthesized during the early minutes after the Big Bang, when conditions prevailed not unlike those associated with the temperature and pressure within the explosion of a hydrogen bomb. We say "must have" because it has not proved possible to model the production of the cosmic relative amounts of these lightest elements inside stars, which is the only locale other than the early universe where the inferred physical conditions are sufficiently severe to permit the formation of new nuclear species. (Chapter 4 mentions some recent efforts by a few cosmologists to explain the formation of *all* the elements inside stars, a view that is not generally accepted.)

The first suggestion that the early hot, dense state of an expanding universe was the site of the formation of the elements was made by the late George Gamow in 1942 in a lecture before the Washington Academy of Sciences (abstracted in its journal) and, in somewhat greater detail, in 1946 in *Physical Review*. The mathematical basis for the model was derived, as already mentioned, from the work of Einstein in 1917 and from subsequent work through the 1940s by Alexander Friedmann, Georges Lemaître, Willem de Sitter, H. P. Robertson, A. G. Walker, R. C. Tolman, and others. The earliest calculations of cosmic nucleosynthesis in such a model were carried out by Alpher with Gamow; our first results were published in 1948 in what became known as the Alpher-Bethe-Gamow or alpha-beta-gamma paper (see chapter 3, for example, and books by Bertotti et al., Hetherington, and Kragh, in the recommended reading list) and were improved and extended by Alpher and Herman; we continued to collaborate on various aspects of this work until Herman died.

The descriptive word *bang* for the origin of the universe at a finite time in the past appeared in a 1928 book by the British astronomer Arthur S. Eddington *The Nature of the Physical Universe* (Cambridge: Cambridge University Press, 1928)—long before the Big Bang model was formally articulated. He was at the time concerned that there seemed only two possibilities for the universe: that it had existed forever, or that it had an origin in the finite past, and, he said, "I simply do not believe that the present order of things started off with a bang. . . . Philosophically, the notion of an abrupt beginning of the present order of Nature is repugnant to me." Remember that this was written before the evidence for an expanding universe had penetrated the literature, and certainly well before the idea of an evolving universe with a hot, dense origin had become current. The term *Big Bang* appears to have been used first by British cosmologist Fred Hoyle (who probably did not remember Eddington's remark) in a 1949 transatlantic radio debate on the BBC with George Gamow about the relative merits of what we then called a "dynamic evolving model." Hoyle was comparing the "Big Bang model," with the "Steady State or Continuous Creation theory," which he had helped develop. Gamow told us that he thought the discussion became heated. Most cosmologists are convinced that the Steady State model is no longer a viable concept. Like it or not, and despite a recent national contest held by the magazine *Sky and Telescope* (announced in the August 1993 issue, and the judges' conclusion in March 1994 that the name should not be changed) to find another name, the expression Big Bang is alive and well and probably too firmly embedded in the popular and scientific literature to be changed.

It is now reasonably clear that the cosmic relative abundances of the heavier chemical elements with masses greater than those of hydrogen, helium, lithium, and perhaps beryllium and boron, up through the element iron, are indeed established in stellar interiors during the normal evolution of stars, as the result of energy-generating thermonuclear fusion reactions. The energy release due to the onset and continued fusion of light elements through helium and up to iron provides radiation pressure that prevents further gravitational collapse of the body of the star. When fusion is no longer possible (i.e., fusion fuel is exhausted), then elements heavier than iron are synthesized in the further evolution of stars that begin to collapse. These stars evolve into white dwarfs, neutron stars, or black holes, depending on how large the initial mass was when the fusion fuel in the core was used up. According to theoretical models of stars, those lighter than about 1.4 solar masses evolve into white dwarfs; stars with masses between 1.4 and about 3 solar masses evolve into neutron stars; and heavier stars probably evolve into black holes.

A white dwarf is a star that has exhausted its nuclear fuel, collapsed, and gone through a series of states in which the gravitational energy released in the collapse as internal energy generates outward pressure that counteracts the inward pressure of gravitational forces. A neutron star has also exhausted its fuel; it explodes during its collapse and leaves a remnant that in turn collapses to a state in which all nuclei are in effect compressed into a dense neutron gas. This gas then exerts what is called *degeneracy pressure* (a quantum mechanical preclusion of available states for particles) which prevents further collapse. Neutron stars have interesting observational properties in that they may become a type of star called a *pulsar,* a spinning star which radiates electromagnetic energy from a restricted area of its surface in a kind of searchlight mode as it loses further energy. As the star rotates, this beam of radiation sweeps through space and, if the alignment is favorable, over the earth, yielding observable periodic pulses of radiation. These pulses can be as rapid as thirty times per second. A black hole is a star that begins with an explosion as it evolves, and develops as a supernova, but the residual star is so massive that its surface gravity prevents the escape of radiant energy, which is why it is called "black." Black holes can, in principle, be observed only in terms of their gravitational effects, magnetic field, and charge, if any. In fact, however, evidence for the existence of black holes is accumulating in the details of the velocity fields and radiation characteristics from *accretion disks,* disks of stars and matter caught up in the strong gravitational field near black holes.

In general, the elemental species heavier than iron are synthesized during the violent infall of matter in massive stars, stars that have exhausted their nuclear fusion fuel, and undergo subsequent explosion. White dwarfs, which were relatively less massive stars, do not explode immediately following the collapse phase. Rather, they accumulate matter on their surfaces from nearby companion stars, and from time to time thermonuclear explosions occur in the puddles on the surface. Modified nuclear species are then distributed into space, where they may be swept up and accreted in the next generation of star formation in the local interstellar medium. Thus, the first generation of stars contains only hydrogen, helium, and a trace of the other light elements; later generations are enriched with small amounts of the heavier elements, which seed interstellar space as ejecta from earlier generations. The luminous output for typical supernovae that evolved quickly goes through a maximum and then dims on a time scale of dozens of days. Such light curves play an important role in current efforts to establish more firmly a distance scale for objects in which these supernovae are observed.

The agreement is surprisingly close between the observed cosmic relative abundances of the light elements and those calculated in a detailed way within the Big Bang model in the process termed primordial nucleosynthesis. Moreover, the theoretical models of heavy-element nucleosynthesis occurring in evolving stars and their subsequent explosions are also in good agreement with observation. These models were first explored in detail in a classical paper in a 1957 *Review of Modern Physics* by Margaret Burbidge, Geoffrey Burbidge, William Fowler, and Fred Hoyle (B^2FH, for short). Modeling the dynamics and nuclear physics of stellar explosions is very difficult, but in recent years much progress has been made. It is satisfying that current observations of such exploding stars as supernovae have verified the general features of the theoretical models.

In calculations of primeval nucleosynthesis, it is necessary to introduce as an initial condition the composition of the universe just prior to its cooling to a temperature low enough for survival of the nuclides formed in nuclear reactions from elementary constituents: that is, sufficiently low that formed nuclei are not broken up by collisions with high-energy photons in a process called photodisintegration. These initial conditions can in turn be calculated from a description of the universe at an earlier time, when it contained only radiation and the elementary constituents of matter. In order to obtain agreement between calculated and observed abundances, it is necessary to assume that during the period prior to primordial nucleosynthesis there were three species of neutrinos present—namely, electron neutrinos, muon neutrinos, and tau neutrinos—and their associated antiparticles.

It is well developed in both theory and observation that there exists an antiparticle for every elementary particle found in nature. For electrons, for example, there exist particles called positrons, the same in every respect as the electron except that the charge is positive rather than negative. And so it is that for protons there exist antiprotons, and for neutrons, antineutrons, and so on. This is certainly at odds with one's intuition. The first published suggestion (1934) was due, it appears, to Vladimir Rojansky, when he was a professor of physics at Union College, who proposed in *Physical Review* that there should be two kinds of matter, which he termed *terrene* and *contraterrene* matter. The basic work on antimatter was contained in the work of the British physicist Paul A. M. Dirac on the theory of electrons. He proposed that there should be a kind of mirror particle for electrons, which would have the same properties as electrons except positive charge: namely, mass and spin would be the same. (The latter is a number characteristic of elementary particles, called a *quantum number*. There are two kinds of particles in nature in terms of this quantum number: fermions, which have spin, and bosons, which do not have spin.) The antiparticle for the electron, named the positron, was discovered by Carl D. Anderson to exist in cosmic rays, particles that come to earth from outer space. Since that discovery, antiparticles have been identified for the other elementary particles, primarily in laboratory observations using very high-energy particle accelerators. The neutron is an interesting exception in that it has no intrinsic charge but does have a magnetic moment—as though a small magnetic field were being generated by a circulating current in the neutron—so there is an antineutron. One fascinating aspect of particles and antiparticles is that should a particle encounter its antiparticle, the pair will annihilate, with the emission of energy in the form of a pair of high-energy photons—that is, gamma rays. All of this has been observed as well as theoretically predicted. The multiplicity of energy states represented by elementary particles and antiparticles is reflected in calculations of the properties of the high-temperature plasma that preceded the onset of the formation of light elements in the thermonuclear furnace of the early radiation-dominated universe.

There is another feature of the prenucleosynthesis era of the early universe. Not only must we deal with the contributions of electrons and positrons, baryons (neutrons and protons) and antibaryons, quarks and gluons (constituents of baryons and antibaryons) but we must say something about the nature of neutrinos and their antiparticles, which should have been present. Again, we digress a bit to describe neutrinos, which surely play an important role in arriving at nucleosynthesis. Some sixty years ago there was study of a class of radioactive nuclei called beta emitters, which decay by emitting an electron (or a positron, for certain nuclei). But measurements of the energy and spin of the emitted particle showed a deficiency: spin was not conserved, as it should be; moreover, there was a distribution of energy of the emitted particles, even taking into account the recoil energy of the parent radioactive nucleus. A hypothesized solution was a new kind of particle, now called the neutrino, carrying spin, as well as the discrepancy in energy, but no mass (or none to speak of), which corrected the discrepancy in the energy spectrum of the emitted particles. Later, the beta emitters that emitted positrons were found, and for these it was proposed that an antineutrino was associated with balancing the reaction.

It now appears that there are at least three types of neutrinos and their antiparticles: the electron neutrino, the muon neutrino, and the tau neutrino. The electron neutrinos, and their antiparticles, are associated with beta decay in general; the muon and tau neutrinos are associated with the decay of other elementary particles. In the original study of prenucleosynthesis calculations done by the authors and James W. Follin Jr., in *Physical Review* in 1953, knowledge of the types of neutrinos was limited. Later calculations of light-element nucleosynthesis suggested that there should be three types. The requirement for three neutrino species was a cosmological prediction, and it has been very satisfying to cosmologists that the existence of the three kinds of neutrinos has since been verified in laboratory observations using very high-energy particle accelerators. The agreement is now frequently cited as an additional piece of evidence for the Big Bang model (although there may be a problem with it, probably having more to do with some lack of precision in the observed and presumed primordial abundance of helium and deuterium than in the neutrino requirement).

The final piece of evidence is called, for convenience, the *three-degree radiation*. Its existence in an unambiguous set of observations was initially reported as a 3.5 kelvin unpolarized isotropic blackbody radiation by Arno A. Penzias and Robert W. Wilson in 1965 at Bell Telephone Laboratories in Holmdel, New Jersey. The temperature of this microwave radiation background has been determined by a polar orbiting satellite called COBE (Cosmic Background Explorer), launched by NASA for the purpose of making measurements related to this background. From accumulated measurements over four years the contemporary value is 2.725 ±0.002 kelvin, based on an extremely precise blackbody radiation spectrum (see chapter 5 and the book by Mather and Boslough cited in the recommended reading list). (The kelvin temperature scale refers to degrees above absolute zero; on the kelvin scale, the freezing point of water—32° Fahrenheit or 0° Celsius—is about 273 kelvin; 68° Fahrenheit is about 293 kelvin.) The concept of blackbody radiation goes back to the end of the nineteenth century and was developed principally by Max Planck in Germany. We now speak of blackbody radiation as exhibiting a *Planck spectrum*.

Again, we digress to discuss the Planck spectrum of blackbody radiation in some detail, since it is one of the main features of the Big Bang cosmological model. Different wavelength portions of the electromagnetic radiation spectrum can be generated in a variety of ways. For example, radio waves are emitted by an electrical conductor carrying an oscillating current; any entity at a temperature greater than 0 kelvin emits thermal radiation; an electric discharge through a gas may cause visible or ultraviolet radiation to be emitted; x-rays are emitted from a metal bombarded by high-speed electrons; radioactive atoms may emit gamma rays; certain substances emit fluorescent radiation when irradiated by appropriate wavelengths of electromagnetic radiation; and all sorts of electromagnetic radiation are emitted from nuclear reactions. In the Planck spectrum, we deal with thermal radiation.

It has been shown by experiment that the rate at which a body emits thermal radiation depends on the temperature and the specific character of the surface. The total radiant energy emitted by a body per second per unit of area is termed the total emissive power. On the other hand, when thermal radiation falls upon a body, some of it is absorbed, some reflected, and some transmitted. The fraction of the total energy of isotropic radiation (i.e., radiation coming equally from all directions) that is absorbed is termed the total absorptivity.

Some substances such as carbon black have a total absorptivity that is almost equal to unity, since virtually all the radiation that is incident upon it is absorbed. Visualize for the moment an ideal substance capable of absorbing all the radiation that is incident upon it. A body of this type is called a blackbody, whose total absorptivity is then equal to unity. An excellent approximation to a blackbody is a cavity whose walls are kept at a fixed temperature and whose communication with the external world is through a hole whose size is very small relative to the overall dimensions of the cavity. A properly designed furnace is an example of such a cavity, and furnaces with small openings are commonly used in the laboratory as sources of blackbody radiation. Note that any radiation entering the small hole, as well as any radiation emitted by the interior walls of the cavity, is in part absorbed and in part diffusely reflected many times by the walls, so that only a very negligible fraction will finally escape through the hole. The very small amount of the radiation that escapes through the hole has a character that is independent of the substance making up the walls. It is called blackbody radiation with the so-called Planck distribution of intensity as a function of the wavelength. Thus, the total emissive power of a blackbody depends on the temperature only. In 1879, Josef Stefan deduced empirically that the total emissive power of a blackbody is proportional to the fourth power of the absolute temperature. In 1884, Ludwig Boltzmann explained this result on theoretical grounds, and the fourth-power temperature dependence has been known ever since as the Stefan-Boltzmann law. Planck's description of the emission of a blackbody came later in the history of physics.

To reiterate, the blackbody radiation, which is in equilibrium with the walls of the cavity, has a total energy that depends on the temperature and on the volume of the container, not on the material making up the walls. It has been shown that the energy of the radiation is directly proportional to the volume of the container, so the ratio of the energy to the volume, or the energy density, is a function of the temperature only. We know from theory and experiment that radiation exerts a pressure, and from electromagnetic theory and observation that the pressure exerted by blackbody radiation on the walls of an enclosure is equal to one-third the energy density.

Another matter of importance to our discussions is the response of a container of blackbody radiation to a change in volume of the container, without any energy being added to or removed from the container. In this case, consider that the blackbody radiation in a cavity with perfectly reflecting walls is in equilibrium (practically speaking there must be at least a speck of matter in the enclosure for the radiation to interact with). If this type of cavity expands, then the expansion is said to be adiabatic, a term which means that there is exchange of energy neither between the radiation and the walls nor with the outside world. The work done on the surroundings comes at the expense of the internal energy of the radiation and the speck of matter, which can be neglected. Throughout the expansion, the radiation is in equilibrium with the matter, and it can be shown that the product of the volume containing the radiation and the third power of the temperature is a constant. This means that if, for example, the volume containing the blackbody radiation were increased by a factor of eight, the equilibrium radiation temperature would be reduced by the cube root of eight, or two—namely, to one-half the original value.

Again, the intensity of the radiation at a given wavelength emerging from the small hole follows a distribution pattern with wavelength that depends only on the temperature of the cavity. This is called the Planck distribution, and it is written down in the appendix—as are approximations to this distribution for long wavelengths, called the Rayleigh-Jeans formula (due to Lord Rayleigh and Sir James Jeans), and for short wavelengths, called the Wien formula (put forth by Wilhelm Wien). There is also a Wien displacement law, which says that the maximum intensity in the Planck distribution is displaced toward longer wavelengths as temperature increases. The Wien displacement formula is useful in calculations, since it in effect states that in the Planck spectrum the wavelength at maximum intensity multiplied by the temperature in degrees absolute, that is, on the kelvin scale, equals a constant whose value is 0.29 centimeter kelvin. Thus, at 2.728 kelvin, the wavelength at maximum intensity is 0.29 (in units centimeter kelvin) divided by 2.728 kelvin, or about 0.1 centimeter, which is in the microwave region of the electromagnetic radiation spectrum. In the cosmic context, it is as though the universe were a cavity filled with radiation at a temperature of 2.728 kelvin. That the universe can be seen as a blackbody cavity is an approximation justified by the overwhelming dependence of the properties of the early universe on its radiation content.

A gas of any kind of particles exerts a pressure against the walls of a containing vessel. Now again, suppose that the container of blackbody radiation undergoes a change in volume. Then work is done in an amount equal to the product of pressure and the volume change; this work represents the utilization of energy stored internally in the gas, in, for example, the kinetic energy of the particles or in the energy of the photons. In the cosmological context, where one models a universe in which the dynamical behavior is controlled by radiation, we consider that the radiation in the otherwise virtually empty container which is the universe is behaving like a gas of photons, doing work against the walls. There is a conceptual problem here: namely, that in an infinite universe there are no identifiable walls.

As we have said, if there is no way for energy to be created in the container or removed from it, any change in volume of the container is called adiabatic, as is the consequent change in volume of the gas. If the container expands, then the gas expands adiabatically at the expense of its internal energy; consequently, it cools. This energy becomes work done and, in laboratory-scale experiments, goes into motion of the walls of the container. The universe is a bit more complex, since as just remarked, there are no walls on which work is done. Therefore, we have a wall or boundary problem. Where does the energy go that is represented by this work? Enrico Fermi once suggested at the end of a colloquium by Alpher that the work done, which is equal to the energy expended by the expanding gas, goes "into the hands of God," which is another way of saying that we do not know and that there may be problems in applying thermodynamic concepts to the universe at large. In technical treatments of this problem of the global nonconservation of energy, many cosmologists conventionally confine their attention to limited, finite segments or volumes of space, so-called comoving volumes. This may not satisfy some readers, but it is a quite usable operational view of the problem.

There has been no accepted physical explanation for this background blackbody radiation other than that it is indeed a fossil of the early Big Bang, a very much cooled (more correctly, redshifted) relic of the radiation that pervaded the universe some hundreds of thousands of years after the Big Bang. Russian cosmologists for a time preferred to use the term *relict radiation,* and the former Soviet Union orbited a satellite called RELIKT to measure it. Our original theoretical studies of the Big Bang model led us to conclude in 1948 that such radiation should exist and should have reached a temperature of about 5 kelvin. This predicted value depended on the then current values of the cosmological physical parameters; using now-current values of the parameters yields the COBE-determined value of 2.725 kelvin, a number with a small experimental error of a few units in the last decimal place. It is amusing in this context that the late famous mathematician Paul Erdös is said to have re-marked that God made two "mistakes": he started the universe with a Big Bang, and then he left the three-degree radiation behind as evidence.

Before leaving this discussion of the background radiation, we feel it is appropriate to comment on Planck, Rayleigh, Jeans, and Wien. Long before there was such a concept as quantum mechanics, Planck proposed that photons of light were quantized—that is, manifested in discrete packets of energy—and suggested his distribution function, based on empirical information. Rayleigh, Jeans, and Wien followed with their approximations. The idea that light is quantized really became fully ac-cepted when Einstein proposed such discrete packets of energy in explain-ing the photoelectric effect: the emission of electrons from solids result-ing from the impact of photons on the surface. This is an example of the observation that physical understanding does not always proceed in a logical and orderly fashion.

What Then Is the Big Bang Model?

The Big Bang model describes a universe that is dynamic and evolv-ing, one that started from an extremely hot and dense state at a finite time in the past, then expanded and cooled to what we now observe. It is based on Einstein's theory of general relativity, from which one can derive a mathematical relation frequently called the Friedmann-Lemaître (or the FLRW) equation, showing how selected physical vari-ables in the universe change with time. Examples of such variables include the separation with time of any pair of points in the universe arbitrarily selected and thereafter tracked, the smeared-out density of matter, and the mass density equivalent of the energy contained as radiation. The radiation density, or its equivalent mass density, can be calculated from knowledge of the temperature, using as the conversion the Stefan-Boltzmann relation. Given a knowledge of the smeared-out total density and the expansion rate of the universe, one can in this model calculate the age of the universe, which is nothing more or less than the elapsed time since the Big Bang event. (See the appendix for the mathematics relevant to this paragraph.)

We can visualize cosmic evolution by using the Friedmann-Lemaître equation to describe the universe equally well forward or backward in time, much as a motion picture film can be run and viewed in either di-rection. Consider, then, running the movie back to one second after the

Big Bang, when the universe had expanded and cooled from almost un-imaginable conditions of high temperature and density to a temperature of 10 to 15 billion kelvin and a density of about half a million times that of water—conditions still difficult to comprehend. This high density was almost entirely due to the mass equivalent of the energy in the electro-magnetic radiation that was present. Call the radiant energy E; then, according to Einstein's famous formula, $E = Mc^2$, energy is equivalent to mass M, with the constant of proportionality being the square of c, the velocity of light. At the corresponding epoch, about one second, the density of matter was just a trace amount, more than a million times less dense than the radiation. This trace amount of matter in the form of neutrons, protons, electrons, and a variety of other elementary particles such as neutrinos had evolved from earlier and extreme conditions of temperature and density; the matter at that time was in the form of more elementary relativistic particles, particles moving at very nearly the ve-locity of light, with a correspondingly very high temperature.

Having returned to a time of about one second after the Big Bang, consider time reversed again to go forward. Before about one second, the temperature and density were so high that the neutrons and protons that collided did not coalesce; they were not able to do so because of very energetic collisions with other particles and radiation. After about a hundred seconds, nuclear reactions could ensue because the collisions were no longer sufficiently energetic to break apart the new nuclear spe-cies formed. Nor was the radiation present sufficiently energetic, in events called photodissociation, to break up the nuclei as they formed. Details of such reactions have been observed in the laboratory in high-energy particle accelerators and, with some difficulty, in thermonuclear bomb explosions. The capture of neutrons by protons represents the first step in the buildup of nuclei, including the heavy isotopes of hydrogen—namely, deuterium and tritium (these each contain one proton but one and two neutrons, respectively)—normal helium and its light isotope helium 3, and a small amount of lithium.

These thermonuclear reactions were terminated by the continuing universal expansion, which reduced the probability of collisions as the particle densities dropped; by the overall cooling in the expansion, which reduced the energy of the collisions; and by the radioactive decay of neutrons into protons, electrons, and neutrinos, which removed neutrons that could otherwise have been involved in various reactions. By the usual definition of half-life, the neutron is a radioactive particle with a half-life of about 12.8 minutes: in a given quantity of free neutrons, half would undergo decay in about 12.8 minutes. This exponential decay process limited the time available for neutrons to be involved in reactions in the element-forming process. All of this nucleosynthesis occurred in a time from about one hundred seconds to about three hundred seconds after the Big Bang.

Following primordial nucleosynthesis, nothing much of interest seems to have occurred during the next 100 thousand to a million years except for expansion and cooling. During this transition epoch the rate of expansion was controlled by the radiation present because of the much higher radiation density in the early universe, and the overall state of the universe was that of a very dilute gaseous plasma containing trapped radiation. Trapped radiation is radiation that is absorbed and re-emitted, or deflected, or scattered by the atoms present at distances comparable to the average separation of the atoms in the plasma. The plasma consisted of stripped hydrogen atoms, which are protons, that is, fully ionized bare nuclei of hydrogen, together with a free electron each and stripped helium atoms (bare helium nuclei, with two positive charges and two free electrons each). The plasma was a fully ionized gas that on the average had no net electrical charge. After three to five hundred million years or so, the temperature had dropped to several thousand degrees, sufficiently low for the free electrons to combine with the hydrogen and helium nuclei to form neutral atoms. The gas then transformed from a plasma state to a neutral gas, a transition process called recombination. The rate of neutralization was finite but short compared with the expansion rate of the universe. This was the first time in the history of the universe that neutral atoms would have appeared.

Several phenomena are associated with this era of recombination. For one thing, as stated earlier, while the gas was still a plasma, radiation could not travel very far before the photons were deflected or scattered by interaction with charged particles, whether stripped nuclei or electrons. Such a system is said to have been opaque; the radiation was effectively trapped in the plasma. When the gas became charge-neutral, when plasma recombination occurred, the radiation could travel more freely. The universe had then become transparent to radiation, and we can say that matter and radiation were effectively decoupled. Moreover, the radiation emerged at this time with the same degree of local and global homogeneity and isotropy that was exhibited by the matter present in the plasma state during the time of last interaction of the radiation with the plasma. A second noteworthy phenomenon was very near to the conditions of recombination when, the model predicts, the nature of the expansion changed over from one controlled by the mass density of radiation to one controlled by the mass density of matter. It is just during this epoch that the density of radiation dropped to a value below that of matter. That this crossover event occurred at all is a reflection of the cosmological model in which the radiation density decreased more rapidly with the cosmic expansion than did the matter density (for more detail, see the appendix).

We reiterate that the decoupling allowed the noninteracting and therefore freely propagating radiation to retain and reflect any small departures from homogeneity and isotropy in the spatial distribution of matter that existed in the universe at the time of decoupling. Just such small departures, observed to be at the level of about 10 parts per million, have been recently reported with considerable excitement from analysis of measurements made over a period of four years by the COBE satellite and by instruments set up for this kind of observation in the Antarctic (flown in high-altitude balloons) and elsewhere. These small departures were hoped for and anticipated by cosmologists to act as the needed seeds for the later gravitational agglomeration of matter into stars, stellar clusters and galaxies and clusters of galaxies, in whichever order these events occurred. Had there been absolute homogeneity, or departures from isotropy significantly smaller than those now observed, it would be difficult to develop a theory that would explain the formation of cosmic structures in the time allowed by the age of the universe since decoupling.

The decoupled radiation has freely expanded, with no further significant interaction with matter, from the recombination era until the present day. Expanding radiation cools adiabatically, and thus the universal blackbody radiation had cooled from several thousand kelvin at the time of decoupling to the 2.725 kelvin background temperature measured by COBE, with the temperature dropping inversely as the scale factor. Note that according to theory, neutrinos—the very light or massless particles mentioned earlier—should also have persisted to the present time, cooling with the expansion, and according to calculations should be pervasive in the universe at energies corresponding to a temperature of about 2 kelvin. Although theoretical calculations suggest there should be several hundred per cubic centimeter, with present technology it does not appear possible to detect such cold neutrinos. One must be cautious in this regard, however, because prior to the measurements of Penzias and

Wilson, the possibility of observing the cosmic microwave background radiation was similarly doubted by most observers. There might also be pervasive gravitational waves, generated by complex motions of matter in the time before recombination occurred, left over from the early universe, their observability again highly unlikely with present technology. Recently it has been suggested that prior to decoupling there were acoustic waves, with a fundamental wavelength corresponding to the dimensions of the universe at the time. Still, we are fortunate in having several well-accepted observed fossils of the early universe, including the cosmic abundance distribution of the light elements and the expanded and cooled blackbody radiation, and the tiny departures from uniformity from the recombination era, whatever their origin.

The minuscule deviations from homogeneity in temperature in the universe at the time of last interaction of radiation with matter, as now observed by COBE, are in fact small local excesses or fluctuations of temperature, which can be interpreted as very tiny fluctuations of the density of matter and radiation. These small deviations, scattered through the cosmos, could then have acted as seeds for the gravitational condensation and growth of the various structures in the cosmos. Until the last few years it was considered a problem of the Big Bang model that the cosmic background radiation appeared to be uniform, to levels of the order of one part in 100 thousand. As stated earlier, cosmologists felt that some type of seed for condensation was required to allow the observed structures in the cosmos to develop in the available time since the Big Bang.

We have not been involved in what have come to be called inflationary models, which are embedded in many current studies of the extremely early universe. The inflationary concept is that for an extraordinarily short time after the Big Bang, the universe expanded at an enormously rapid rate; then, at some very early point in time, the expansion smoothly changed over in its behavior to that of the standard Big Bang model. This assumed rapid expansion, not yet explicitly verified, has some interesting consequences for the development of anisotropies.

Unfortunately, our knowledge of the observed early anisotropies—these seeds or matter-density fluctuations and, in particular, their size distribution—is not yet sufficiently precise to enable us to select among the several theoretical models proposed for the formation and evolution of subsequent structure in the universe. Perhaps detailed understanding will become available with increased activity in observing these fluctuations at different size scales. At least for now, the scientific question is no longer how structure developed without seeds to start condensations but, rather, what the details of the process of structure formation from the observed anisotropies might be. Theories of the formation of structure are very much in a state of flux, and we shall leave to others the task of describing such work. For example, as we have already intimated, we do not know with any certainty whether structures formed "top down," with galaxies and clusters of galaxies appearing first, followed by star formation, or "bottom up," with stars appearing first, followed by agglomeration into galaxies and clusters.

The Very Early Universe

We complete this overview of the Big Bang by returning to the very earliest stages of the model, for which we promised no more than the briefest of descriptions. It is now generally accepted by the proponents of the Big Bang that the universe evolved from an early hot, dense state some 15 to 20 billion years ago. The age estimate is still in some dispute as this book is written, because in the model it depends on knowledge of the expansion rate of the universe as well as the sum total of the densities of matter and radiant energy. These measurements are an active field of research, and there is movement toward consensus but as yet no convergence of various measurements to consistent values. We suggest an age of 15 to 20 billion years for present purposes, with 15 being more likely.

At the time of writing, the several schools of interpretation of measurements of the expansion rate have moved closer together. A few years ago the difference was about a factor of two; now, their probable errors of measurement almost overlap, depending on what one takes to be the universal density of matter. It is anticipated that with the right choice of density, the several approaches will arrive at an age for the universe that is greater than the age of any of its constituents (recall that the age proposed by one of the approaches is possibly less than the derived age of the oldest clusters of galaxies). We refer ahead to an illustration in chapter 3 that shows the observations of expansion rate plotted against the year of observation (figure 3.3). These are all indeed innovative and difficult observations, but the results are surely converging!

The initial hot, dense state can be viewed in a number of ways. A strict and perhaps naive application of Einstein's general relativity theory to the beginning of the universe, which provides the mathematical framework for the Big Bang model, suggests that there was an initial event that can be described as a singularity. In physics and mathematics the existence of a singularity is a situation in which things "blow up"; in the Big Bang model one says that the temperature and density were "infinite" at the origin. You may remember that your mathematics teachers cautioned you never to divide by zero, because this operation led to an indeterminate result—confusing as it was, some may even have called the result a singularity.

Steven Hawking and Roger Penrose years ago showed on a theoretical basis that the existence of a singularity was a logical result of modeling the early universe or black holes using the framework of general relativity theory. In the case of the black hole, this analysis has a fraction whose denominator goes to zero under certain conditions. The analysis leads to the relativistic description of the gravitational field around a point mass (or a spherically symmetric nonrotating mass); the singularity then suggests the definition of a black hole. Yet such a result is disquieting to scientists, who generally believe that nature does not tolerate singularities. Most would argue that if the physics of the situation were known well enough, there would indeed be extreme conditions in a physical sense but not a strictly defined singularity. At the very least, it appears reasonable to believe that quantum phenomena are deeply involved in the physics of the very earliest universe, rather than simply the general relativistic description of the behavior of matter and energy under extreme conditions. Quantum physics is usually used to describe the behavior of matter on a very small scale: that is, in discussing atoms and elementary particles. There has yet been no successful marriage of quantum mechanics and general relativity—a major current effort of "string" theorists (see, for example, the book by Brian Greene, in the recommended reading list), a body of work which we shall not deal with here.

Those who are trying to treat models of the very earliest universe in a quantitative way suggest that with some stupendous extrapolations from current understanding of high-energy particles and fields, one can begin to look at the physics of the earliest epochs—that is, after an almost unimaginably short time following the Big Bang event: namely, about 10^{-43} seconds. The choice of this so-called *Planck time* is by no means arbitrary; it can be rationalized in several ways. One way involves a procedure known as dimensional analysis, which leads to a set of quantities known as the Planck or the Planck-Ginzburg numbers (which include the Planck time), after the scientists who first suggested their importance in physics. These quantities are units of measurement of mass, length, and time in which the gram, the centimeter, and the second are replaced by quantities obtained by combining appropriate powers of the velocity of light, the Planck constant (which is the energy of a thermal radiation photon per unit frequency), and the constant of gravitation (which appears in Newton's law of gravitational attraction between two masses). The units are described in the appendix. (See also chapter 9).

At this extremely early Planck time all that is contained in the present potentially visible universe (i.e., within a sphere whose radius is given by the velocity of light times the Planck age of the universe) was confined or generated in some way in a region that was exceedingly small compared with the size of the atomic nucleus. Note that 10^{-43} seconds times the velocity of light equals 3×10^{-33} centimeters, one of the Planck units, which is much smaller than any known elementary particle. For comparison, the radius of the hydrogen nucleus is about 10^{-13} centimeters. The smallest piece of matter with which physics deals is the electron, which is believed to have zero dimension—to be a true point particle—which is strange indeed.

One concept applied to this extraordinarily early regime is that it was initially a vacuum state containing a tremendous amount of negative energy, the potential energy associated with gravitational attraction being negative, compared with the usual positive kinetic energy, the energy associated with the motion of particles. It is conjectured that a statistical fluctuation, a momentary local departure from uniformity, led to the appearance of the constituents of elementary particles as we know them, as well as energy in the form of radiation and exotic fields, all emanating from the transformation of the negative potential energy. The transformation is assumed to have occurred in such a way as to conform to the laws of conservation of energy, momentum, and electrical charge, although the transformation must have developed a large amount of entropy—a measure of the disorder in the system. As the population of particles and fields in the fluctuation grew, the region of space embodied in the fluctuation—which even then contained all that now exists in the observable universe in the form of matter or energy—underwent a violent expansion, with dimensions increasing by an enormous number of factors of ten, perhaps as much as 10^{30} to 10^{50} in the very brief period "stretching" from 10^{-43} seconds to 10^{-30} seconds. At this time, it is supposed, the inflation ceased, and the model subsequently transformed into the conventional Big Bang model.

Such an inflationary period would produce several consequences. First, as already mentioned, it would have been during this period that the entropy of the universe (i.e., the level of order) would have been established. Another consequence is the possible occurrence of processes known to exist in laboratory studies of particle physics, wherein despite energy considerations one would expect the particles of matter as we know them to predominate over the equivalent particles of antimatter, leaving a cosmos in which matter abundance predominates by at least a factor of 10 million over antimatter abundance; this consequence is associated with "symmetry-breaking" in certain types of elementary particle reactions. These reactions can proceed "forward," forming particles, and, in the "opposite" direction, forming antiparticles. If the forward and opposite rates are not equal, then there is symmetry-breaking. A third consequence is that the temperature of this very early state would have been uniform over the entire nascent cosmos. Fourth, extremely small departures from temperature (and therefore density) uniformity, created by quantum fluctuations during this period, would have been amplified by the inflation, would have survived to the recombination or decoupling era, and would now show up as the tiny anisotropies in the cosmic microwave background radiation.

As already mentioned, recent data on the anisotropy considered as a function of the solid angle of observation (the footprints of the observing beam on the surface of last scattering) suggest that the temperature and density fluctuations can be considered as acoustic waves traversing the universe, with short wavelengths being damped quickly and long wavelengths surviving. In fact, the data suggest that the principal surviving wavelength has a value corresponding to the size of the universe at the time of decoupling. Finally, the region of expansion, growing at many times the speed of light, would now be considerably larger than what we consider an observable horizon.

It is interesting that the times for various events in the classical hot Big Bang model separate nicely and in several ways from the much more speculative modeling of the extremely early universe, as in models of the inflationary regime, which consider the physics during an extraordinarily short period after the Big Bang event. From approximately a millionth of a second (a microsecond) following the initial event, however, there seems to be little doubt about the classical model.

2

Cosmology before
the Big Bang Model

For millennia, *Homo sapiens* has constructed scenarios about the universe in attempts to understand whatever observations could be made, to seek reasons for the existence of the observed universe, and to make some sense of the existence of humankind. There is little doubt that this yearning for understanding has been the basic driving force for the many religions that have developed over the years. Most of the world's religions have accepted the idea that some extramundane entity either started the universe off and still controls it, or started it but has kept hands off since then. Various religions have introduced more or less personal connections with this entity. Most adherents to specific religions find it easier and more fulfilling to ascribe the wonder and mystery of the universe to the initial or continued actions of such an entity rather than to regard the unfinished task of understanding as an exciting search in and of itself. Most formal religions satisfy the apparently intrinsic need to understand why *Homo sapiens* exists at all, and usually promise some manner of existence after death which will be far superior to what is currently being experienced.

If one's faith in these matters is absolute, then there may be no reason to seek further understanding of the observed universe or of the place of *Homo sapiens* in the scheme of things. Complete dependence on religious teachings is not sufficient for everyone, however, for many people seek within their religious faith an understanding consonant with the increasingly detailed and insightful observations of universal phenomena being made by scientists. Over the centuries, humankind has expanded the boundaries of knowledge, and the 20th century appears to have been increasingly rich with such expanded understanding. Yet the organization of observations in the search for ultimate meaning has not succeeded fully, and thus some find it necessary to appeal to the mystery of a creative entity operating outside the observable universe, with various levels of intercession in the affairs of the universe and in the affairs of humankind.

Many cosmologists (this is a personal observation, for we have not done any scientific opinion survey) feel that "religious exaltation" lies in the search for understanding, and they would argue that this search may be the most religious activity of all. Moreover, observational and theoretical skills in cosmology have increased remarkably over the years and particularly in the 20th century. Hardly a day goes by when there is not yet another revelation of the wonders of the universe in which we live, thanks to such instruments as the Hubble Space Telescope, the Chandra X-Ray Telescope and other astronomical satellites, and to new generations of extremely large-aperture and technically sophisticated ground-based telescopes. These telescopes have bypassed the obscuration of our turbulent atmosphere either by going outside of it or by using techniques called adaptive optics, in which the telescope mirrors are deformed continuously to correct for atmospheric distortions, which are revealed either by ground-based laser light reflections from the distortions or by the distortions of light from other reference stars.

Until the recent past, our views of the universe were largely limited to wavelengths of light in the visible spectrum. Life on earth appears to have evolved in such a way as to concentrate its visual abilities in the visible spectrum of light as emitted by the sun and modified by the earth's atmosphere. Some terrestrial organisms have unusual abilities to see light that is either bluer or redder than that accessible to human vision, and some life forms appear to be able to detect polarization of light or to use the earth's magnetic field as a guide to navigation.

But now, new capabilities of scientific instrumentation suggest to cosmologists that humans are on the threshold of a new level of understanding. New astronomical satellite instruments are observing cosmic phenomena at wavelengths, such as those in the x-ray and gamma ray range, which were previously inaccessible from the ground because they are absorbed by the atmosphere. Moreover, ground and satellite-borne instruments study the universe at a variety of radio frequencies and at infrared frequencies that certainly are not accessible to human vision.

Phenomena revealed in the last four decades of the 20th century include an immense variety: many kinds of interstellar molecules, interstellar masers, pulsars, quasars, colliding galaxies, black holes, cosmic background radiation, and more. Surely, as scientists continue to grow more skilled at probing the early universe, we need not cover up our current inability to solve a set of mathematical equations describing the cosmos with the introduction of an extramundane entity responsible for it all. We may never be able to answer final questions, but the thrill may well be in the hunting. In any event, our viewpoint in this book is that cosmology is concerned with the "how" rather than the "why."

Before the 20th Century

What can we learn from the scientists and philosophers of old who did the best they could with very limited observational and theoretical capabilities? Until Galileo, they were limited to naked-eye observations plus some simple instruments to measure relative position, angles, and such. In the 16th century Galileo turned his first rudimentary telescope toward the heavens and began to unravel the mysteries of the solar system. Soon the great Isaac Newton provided the first theoretical framework for understanding the behavior of the cosmos. For what it is worth, one might mention that both Newton and Charles Darwin, whose theoretical and observational work underlies much of modern science, were also clergymen, and the problems Galileo had with the Catholic church's refusal to accept his observations and deductions are well known.

We should not look down our noses at the natural philosophers of the past for their efforts to rationalize the very limited observations at their disposal. Scientists today are, after all, making do with the observations at *their* disposal and using such theories as have been developed thus far in order to model the observational universe. What will pertain a century hence is anybody's guess, but our understanding then will surely build on and probably far exceed what is known now. What we cannot say with any certainty is that there will or will not be a startling new observation that throws a completely new light on the cosmos, or a new theoretical development that sorts out observation in a new and hitherto unsuspected modality. The history of the recent past suggests that both are probable.

In very early times, myths were the basis of understanding the cosmos. The scenarios of yore began with "once upon a time" or with "in the beginning" in the tradition of storytelling. There was emphasis upon the craftsmanship of some extramundane entity that created the universe. Impressed with the renewal of life and the mystery of death, ancient storytellers invoked the concepts of seeds, of eggs, and of birth and the nonfinality of death, all under the guidance of the creative powers of some entity exercising authoritative command over the residents of earth and providing for the edification of *Homo sapiens* the majestic panoply of objects in the heavens (see our later discussion in chapter 9 of the anthropic cosmological principle).

There were some early natural philosophers, however, whose ideas went beyond myths and who began to build a basis in scientific process for understanding the cosmos. We will mention only a few, since we are emphasizing the history of modern cosmology. One was Eudoxus of Cnidus (408–355 B.C.), who calculated the length of the solar year, suggested calendar reforms that were later adopted by Julius Caesar, and proposed that the movement of planets reflected their location on a number of concentric spheres. The Greek philosopher Aristotle (384–322 B.C.) basically adopted the world view of Eudoxus but added the notion of a universe of planets filling space, against a backdrop of fixed luminous objects, and wheeling about for the visual pleasure of human observers. Ideas about these planets were largely qualitative, but a considerable interest in systematic observations was already developing. These observations led to the concept of a finite, earth-centered universe containing planets obeying the laws of geometry as developed by Euclid at about the same time.

A third great astronomer of the ancient world was Ptolemy (Claudius Ptolemaeus, active about A.D. 127–151). He visualized Earth in the form of a globe, stationary in the center of the universe, about which the Sun, Moon, and stars moved in circular orbits at a uniform speed. The cosmic composition then was taken to be earth, water, air, fire, and ether (the last a euphemism for empty space). Planets moved in *epicycles*; an epicycle is a small circle whose center is carried along as the small circle rolls on the circumference of a larger circle. The idea of epicycles persisted through the time of Copernicus, who found it necessary to invoke a considerable number of them to rationalize observed planetary motions. In retrospect, it is useful to remember that several planets were unknown at the time, reducing the number of epicycles needed.

The Aristotelian view survived as the favorite model of the cosmos until the time of Nicholas Copernicus (Nikolai Koperniki), a Polish scientist who lived from 1473 to 1543. (Alpher has a very minor but personal sense of continuity, having visited a university named for Copernicus in Torun, Poland, and having seen his home, carefully restored, where cooking in the kitchen was done on the floor under a chimney). Copernicus introduced a number of new and relatively modern ideas that revolutionized our concepts of the universe. One of the most important was that the universe was populated with stars like our Sun, spread through a vast space. Later, a synthesis of the properties of the solar system, based on Newton's law of gravitation, was the start of mathematical and physical modeling of cosmic phenomena. Another and most important change associated with the Copernican revolution in astronomy was that Copernicus switched our views from an Earth-centered to a Sun-centered universe.

We should not fail to mention the stories of Giordano Bruno (1548–1600) and Thomas Digges (1546–1596). Bruno was burned at the stake by the Inquisition for his radical thesis, based on no evidence, that "our" world was but one of an infinite number of similar worlds. Digges, again with no evidence, opined that the ideas of Copernicus could lead to the possibility of an infinite universe of stars. We do not recall that Digges was in any way involved with the Inquisition; on the contrary, his writings were enormously successful and served to bring the ideas of Copernicus to the attention of the educated public of the time.

With these hints from a few philosophers and early astronomers, we jump to Thomas Wright of Durham (1711–1786), who proposed that the Sun and the solar system were embedded in a disklike finite structure of stars—what we now call the Milky Way. The philosopher Immanuel Kant (1724–1804) built on Wright's ideas to suggest an infinite universe made up of an unlimited number of galaxies like the Milky Way. The early astronomer Sir John Herschel (1792–1871) agreed.

We have only skimmed the surface of the history of cosmology. By the end of the 19th century it was generally accepted that the Sun and the Solar System were resident in a vast assemblage of stars called the Milky Way, which represented the entire visible universe. The Milky Way was seen to contain a huge number of stars and a large number of relatively faint nebulosities. A 20th-century breakthrough identified many of these nebulosities as other "island universes," similar to but not in the Milky Way and at great distances from it. There was in fact a great and historic debate, much publicized at the time (1920), among the three astronomers Heber D. Curtis, Harlow Shapley, and Knut Lundmark on the nature of these nebulosities. The outcome of the debate was by and large in favor of the extragalactic nature of many of these objects, although in time a number of them turned out to be either gas clouds or else clusters of stars in our own galaxy.

The observational breakthrough to the realm of extragalactic nebulae is usually attributed to Edwin P. Hubble, who by 1924 had resolved stars in a number of objects such as the Magellanic Clouds, which are now known to be a small group of galaxies at a distance of about 170,000 light years and identified as members of the Local Group that includes the Milky Way. The Local Group contains some 30 galaxies that are relatively close to the Milky Way and are gravitationally influenced by one another. Just recently, a dwarf galaxy was discovered which is a member of this group. We now know that most galaxies in the cosmos occur in groups or clusters, with as many as a thousand members, bound by their mutual gravitational attraction. Hubble's early breakthrough depended on the low luminosities of some types of stars in distant galaxies that appeared similar to nearby objects, namely, he identified Cepheid variable stars in the Magellanic Clouds, still the most important distance indicator today. In 1929, he published his most important result: namely, the velocity-distance relation now called Hubble's law, which established the phenomenon of the expansion of the universe. Before that was known and accepted, however, the simple attribution of an enormous scale to the universe, as well as its apparent homogeneity and isotropy, had led to the first modeling by Albert Einstein.

At the end of the 19th century, an idea was introduced by Ernst Mach (1838–1916), an Austrian physicist, which Einstein considered important to the theory of relativity. He paraphrased Mach's principle, as it has come to be known, as a statement that the geometrical properties of space-time are determined by the distribution of matter and energy in the universe. In relativity theory we can visualize such properties in terms of the natural trajectory, called a geodesic, followed by a test body moving in curved space-time. For example, the natural path of a planet in the gravitational field of such a central gravitating point mass as the Sun is a very nearly circular orbit. This orbit defines the geometry of space-time near the Sun, as defined in turn by the gravitational properties of the central point mass.

Mach's original version of this idea was that the inertial properties of a body, which determined the acceleration of the body in response to gravitational forces acting on it, had to be based on the gravitational interaction of this body with all other bodies in the universe. Thus, there could be no absolute rotations, velocities, or accelerations. For example, any body assuming a modified shape as the result of a rotation, such as the free surface of water in a bucket being swung around, takes on a surface paraboloidal shape as the result of the presence of and interaction with other material bodies in the cosmos. In other words, the velocity and acceleration of a test particle must be considered in a frame of reference established by the presence of other bodies in the space-time of the interaction.

From time to time there have been articles in the scientific literature on whether Einstein's general theory of relativity, which underlies virtually all contemporary modeling of the universe, does in fact encompass Mach's principle. In general, the question is not answered, but it is not central to considerations of relativity theory. Remember that inertial forces, an easily visualized example being the centripetal force keeping a rotating body on its course (the force acting along a string attached to a ball being swung around), give rise, according to Newton, to an "absolute space." Mach rejected the notion of such a space, arguing that such accelerations were in a frame of reference made up of all the gravitating bodies in the universe. (Interestingly, this view was espoused in a prescient essay titled "Eureka" published in 1848, by the American writer Edgar Allan Poe, who was interested in these matters but not technically trained.) Einstein hoped to encompass Mach's principle in his theory, but, in fact, in general relativity he introduced inertial frames, or "freely falling coordinate frames," in which motions are completely unaffected by either nearby or distant gravitating bodies. A classical example is an inertial frame fixed to the surface of the Earth, from which one cannot detect the gravitational effect of the Sun. It is of course required that there be no gradients of gravitational force in such a frame; that is, it must be free of tidal effects. In the end it appears that Einstein was not certain about Mach's principle, particularly since one can construct alternative models of the expanding universe that are devoid of matter. Willem de Sitter (1872–1934) formulated such models based on general relativity (see appendix). We will say more in chapter 5 about the fact that the cosmic microwave background radiation, which originated in an early region of space-time, does in fact provide a kind of absolute frame of reference for each observer.

The history of cosmology prior to the 20th century displays an abundance of speculation and a dearth of the kind of observational and theoretical capabilities necessary to provide a sound basis for speculations and for modeling. The Milky Way was thought by most to represent the entire observable universe; distances were small by present standards, although large by solar system standards; the universe was thought to be static and forever unchanging; there was no glimmer of the role to be played by nuclear physics in understanding the generation of energy in celestial bodies and the formation of the elements in the early universe and in stars. And in general, when at the limits of physical understanding, recourse was had to the activities of some extramundane entity.

Breakthroughs came rapidly in the 20th century—at least it seems so, although we may be suffering a kind of myopia from living and doing research in that century. A major contribution to the establishment of distance scales in the universe came with the discovery of a relationship between the periodicity in the light output of certain types of stars and their absolute brightness. These stars were called Cepheid variables because of their first observation and association with the constellation Cepheus. This was the work of Henrietta Leavitt at Harvard, supervised by Harlow Shapley, in the 1910s. Absolute brightness was established by first determining the distances to nearby stars by triangulation: measuring the angular position of stars from positions at opposite ends of Earth's orbit about the Sun and then applying simple trigonometry. Given the distance between opposite points of the orbit and the angles of triangulation, and since the intensity of light from a source diminishes with the inverse square of the distance, it was then possible to calculate the absolute luminosity (i.e., the absolute brightness) of the star from the observed apparent luminosity (some corrections had to be applied because of dust obscuration in the Milky Way). In this way a relationship between Cepheid variable periods and their peak luminosity was developed. Then if one observed a Cepheid at an unknown distance and could measure its light-output period as well as its apparent luminosity, one could determine the distance. The existence of Cepheids was a key in Hubble's great discoveries of the distance of galaxies and of the expansion.

To reiterate, the Cepheid period–luminosity relation was and remains very important in that it is the first step in determining cosmological distances, and even today, eight decades later, astronomers search for Cepheids in very distant galaxies in order to determine their distances. The relation has been improved over the years; thus, it now appears that there are two and perhaps more types of Cepheid variables, distinguished by the relative abundance of elements whose light output determines the luminosity of the stars. It is probably fair to say that astronomers are very dependent on Cepheids for getting at cosmological distances, although in recent years it appears that, for example, supernovae (certain kinds of massive stars that explode at the end of their lifetimes) provide a second approach to distance determination, useful to greater distances. We shall say more of this in chapter 3, as well as mentioning some other approaches to distance determination. A major breakthrough has been provided in recent years by measurements made aboard a satellite called HIPPARCOS. Working outside the earth's atmosphere, it has determined with exquisite precision the distance to literally thousands of stars by the classical method of triangulation and has been responsible for establishing new standards for the distances of Cepheid variables.

Special and General Theories of Relativity, and the Original Einstein Static Model

In the first and second decades of the 20th century, Einstein published his epochal papers on special and general relativity. Let us attempt to summarize the impact of these works on cosmology. In special relativity, Einstein questioned the adequacy of classical notions of using measuring rods and clocks to determine spatial and temporal coordinates of events in the physical world, should such events occur in a frame of reference moving with a constant speed at an appreciable fraction of the velocity of light with respect to another frame of reference. This led to the now well-known phenomena of Lorentz-Fitzgerald spatial contraction of rods and time dilation of the rate of clocks. Hendrik A. Lorentz (1853–1928) and George F. Fitzgerald (1851–1901) had earlier been led to such considerations, but Einstein developed a synthesis in his special theory of relativity. An important way to understand the special theory is to observe that the mechanism for comparing rods and/or clocks in reference frames moving at a constant speed one with respect to the other is through the transmission of light signals between the frames, where the velocity of light is not only constant but also independent of the velocity of the source of light or the receiving sensor.

Although much about the theory is counterintuitive, there is no question as to the validity of special relativity. It has been verified repeatedly in a variety of experiments. The motion of particles in high-energy particle accelerators can be understood only in the context of the special theory, for example, and the mechanics of the collisions of high-energy particles can be understood only relativistically. One consequence of the special theory is the famous relationship $E = Mc^2$, which states that the energy and mass of an object are basically the same. The constant of proportionality in this equation is the square of the velocity of light. The principal significance of c^2 here is that it provides that both sides of the equation have the same physical units.

Recall that the inertial mass of a body is a measure of the acceleration of that body in response to an applied force. One of the fundamental assumptions of relativity theory is that there is no difference between inertial mass and gravitational mass; this is known as the equivalence principle. Recent experiments have verified this assumption with high precision: namely, to one part in 10^{13}.

The equation $E = Mc^2$ has achieved popular fame because of its importance in nuclear energy. When a nucleus is split in the fission process, internal energy involved in holding the nucleus together is released and ends up principally as kinetic energy of the fission fragments. The difference in mass between the original nucleus and the sum of masses and energy of the fragments is known as the mass defect of the nucleus. In fusion reactions, on the other hand, such as those that power the Sun and other stars, four hydrogen nuclei are fused at high temperature and density to form a helium nucleus, with the difference in mass between one helium nucleus and four separate hydrogen nuclei basically appearing as energy in the form of electromagnetic radiation and the kinetic energy of the consolidated nucleus. A famous curve of binding energy (see figure 7.1) shows that nature provides a net energy output if nuclei lighter than iron undergo fusion or if nuclei heavier than iron are subject to fission.

We have emphasized that one invokes the special theory of relativity when the reference frames of the observer and the observed are moving at a constant velocity with respect to one another. What now if the two frames are accelerating with respect to one another? An alternative and equivalent question is, what now if gravity is present, with its consequent acceleration of test bodies? This question led Einstein to the general theory of relativity.

What are the essential features of the general theory? For the present discussion, we note that the theory hypothesizes a law above other laws: namely, that we must structure our mathematical descriptions of the laws of nature in such a way that the consequences of the description do not depend in any essential way on the frame of reference chosen. That is, a mathematical description or model of some phenomena in one frame of reference will be the same and produce the same results in any and all other frames of reference. In short, one finds that all specifications of space and time coordinates are relative, and that physics should be the same and invariant, regardless of the observer's state of motion and frame of reference. A consequence of these views is the already mentioned principle of equivalence, which states that there is no difference between inertial and gravitational mass. To say this differently, an observer in a closed box that is being accelerated cannot distinguish what is happening from what the observer would experience in a box that is a freely falling inertial frame, which we have already described.

In a universe governed strictly by Newton's laws, the natural trajectory of a test body far from any sources of gravitational field is a Euclidean straight line. In a general relativistic situation, the trajectory is a curved path whose departure from a straight line at every point of the path is the local curvature of space-time, reflecting local accelerations (or local effects of gravity). In the universe at large, the trajectory of a test body depends on the local net gravitational forces acting at each point; the net effect is that one has replaced gravity by geometry.

A consequence is that one can construct what is called a *line element*, a mathematical expression of the square of a small distance or interval measured along the trajectory. The expression contains statements of the spatial and temporal coordinates at points along the trajectory, which in turn contains explicit dependence on the quantity and distribution of masses elsewhere, masses that give rise to the gravitational effects (space-time curvature) at points along the trajectory.

In cosmology, for a universe that is the same everywhere and in all directions, one can construct what is now called the Friedmann-Lemaître-Robertson-Walker line element, in which the square of an infinitesimal interval along a trajectory depends on the mean density of matter and energy in the universe were they smeared out to a uniform space-time background or matrix. Given this line element, and an equation of state for the background material treated as an ideal gas—that is, homogeneous, isotropic, and not supporting any shears. Shear is most readily described as the ability of a moving fluid to support relative motion in neighboring layers of the fluid. In flow next to a wall, lack of shear means the fluid has the same velocity as in the free flow. From equations relating the response in the density and volume of the gas to changes in pressure, one can derive what has come to be called the Friedmann-Lemaître (FLRW) equation. This equation relates the rate of change with time of any arbitrarily chosen separation of two points in the universe—a so-called *proper distance*—to the mean density of the universe. The most general form of the equation contains a quantity that has been called the *cosmological constant*, which was first discussed by Einstein. He was disturbed that in this equation the time rate of change of separation was not zero. It was his belief, as well as the overwhelming belief of other scientists at the time (about 1917), that the universe was static and unchanging. Modifying the equation to reflect this view required that the constant which he introduced in the equation not be zero. Einstein adjusted the magnitude of this cosmological constant, or Λ, which played the role of a repulsive force, to counteract the attractive force of gravity and stabilize the model universe. In later years, this model came to be called the Einstein-de Sitter model, as de Sitter developed a model of a universe devoid of matter but with the same formal description. The cosmological constant plays an important role, as does the de Sitter model itself, in considerations of an inflationary start to the Big Bang model, as well as in recent observations of a possible acceleration of the expansion.

This first contemporary cosmological model, despite Einstein's attempt to force it to comply with a static universe, was the key to subsequent formulation of expanding or contracting models. Some of the mathematical structure of these considerations is given in the appendix.

3

Development of the
Current Big Bang Model

The origin of modern mathematical modeling of the universe is usually attributed to Einstein. After publication of his general theory of relativity, Einstein developed the first model of the cosmos with large-scale properties as he understood them to be. This was published in 1917 in a paper called "Cosmological Considerations in the General Theory of Relativity." Among the assumptions in this model, the most important to Einstein was that on a scale spanning its major observational features, the universe was homogeneous and isotropic (the same features would be observed wherever one might be located in the cosmos and would appear to be the same in all directions one might look). Moreover, he assumed that the smearing-out of the matter contained in all the visible structures, averaged over the universe, would lead to a single density of matter as a description of the contents of the universe, with the properties of this material content being those of an ideal gas, and with an *equation of state* (a mathematical relation between density and pressure) described by the basic laws of physics for dilute ideal gases. An ideal gas is homogeneous, isotropic, and free of viscosity (i.e., inviscid), so that it will not respond to shear forces. This equation of state, together with an equation describing the conservation of energy and Einstein's basic field equations relating the curvature of space-time to the content of energy (including mass) and momentum, constituted the model's representation of the universe.

The conventional wisdom when Einstein first considered cosmological modeling was that the universe was static (Hubble and colleagues had not yet made their great discovery of the expansion of the universe) and that the peculiar velocities (random motions) of objects in the heavens were all small compared to the velocity of light. To account for this, and in general to force overall static behavior in his mathematical model, Einstein added the cosmological constant to his field equation. The term did not destroy a major feature of his relativistic model, namely, the preservation of covariance, which is to say that the inherent physics of the situation was not affected by the choice of coordinate system in model descriptions. Einstein's concluding statement in his paper said of the introduced cosmological constant, "That term is necessary only for the purpose of making possible a quasi-static distribution of matter, as required by the fact of the small velocities of the stars." He recognized that without this term his model was unstable: that is, it would not be static but would show compression or dilation as time progressed.

Although this added feature did not damage the mathematics particularly, Einstein apparently felt somewhat embarrassed when, in the late 1920s, Hubble, Humason, and Slipher published their observational evidence for a general expansion of the universe; he is said to have called the introduction of the cosmological constant his "greatest blunder." Einstein may have sold himself short. There is a current resurrection of this constant by some scientists for several reasons. For one thing, it makes it possible to reconcile the age of the universe with the age of structures contained in the universe. For another, some cosmologists studying the phenomena of inflation in the universe introduce the cosmological constant to describe a form of negative energy derived from a vacuum energy density that drives the inflation. We will say more about inflation in chapter 6. There is a brief discussion of the Einstein model, including the cosmological constant, in the appendix.

Relative Abundance of the Elements

In the early years of the 20th century, physical chemists—Guiseppe Oddo in 1914, and William D. Harkins in 1917—working with the rudimentary knowledge then available of the relative abundance of the chemical elements, concluded that these abundances must reflect the nuclear properties of the elements rather than their chemical properties. The chemical properties of atoms are determined by the distribution and properties of the electrons associated with the atom, which are in turn conditioned by the electrical charge of the nucleus. As we now know, nuclei contain tightly coupled neutrons and protons, but it is the number of protons (called the atomic number) which determines the number of electrons required around the nucleus to neutralize the nuclear charge and yield a neutral atom. The nuclear properties are involved in nuclear reactions, which appear to determine the abundances of nuclei; the electronic structure determines the behavior of atoms in chemical reactions. Moreover, over the next several decades, determinations of the relative abundances increased in number and scope, including those of elements in the Earth's crust, the atmospheres of the Sun and other stars, meteorites, and the stellar interiors, as inferred from modeling. All of this culminated in the late 1930s with the work of a Swiss-born geochemist, Victor M. Goldschmidt, who constructed a table of the abundance of the chemical elements. His table became accepted as giving the major features of the cosmic abundances throughout the universe.

The observed abundances were thought to reflect the extreme conditions in the locale in which the elements were synthesized, rather than the relatively mundane chemistry of the sites in which the elements were found. This was a profound step forward, indicating that the elements must have been formed in rather extreme physical conditions such as the interior of stars, or, as first suggested by Lemaître and Gamow, in some early configuration of the universe. All of this follows from the fact that nuclear reactions, which alter the nuclei, involve and require energies millions of times larger than the energies involved in chemical reactions. Thus, the elements came to be recognized as possibly a set of fossil remains. Understanding their relative abundance might then be expected to give some clues as to the site of their formation—as indeed it has.

Studies of the composition of the universe and of the stars, dust, and gas contained therein continue apace. The work of expanding and refining the data is much more than a cottage industry among astronomers and geochemists, for improved data give further clues both to phenomenology in the early universe and in interstellar space and to the formation and evolution of stars. Revisions of the data in the late 1940s and early 1950s by Harrison Brown, Harold C. Urey, Hans E. Suess, and by many others since have been illuminating. In particular, increasing attention is being paid to the abundance of the lightest elements, since it is now almost certain that hydrogen, helium, and lithium, and their isotopes were formed by thermonuclear processes in the early stages of a hot, dense expanding universe which has evolved into what we see now. All the other elements in the periodic table—from beryllium, boron, carbon, nitrogen, and oxygen to the heaviest elements, such as thorium and uranium—are formed in stellar interiors. They are then released into interstellar space in the violent outburst accompanying the death of stars that exhaust their source of energy: namely, nuclear fusion in their cores, energy which in radiant form had provided the radiation pressure as a countervailing force to the gas pressure associated with the gravitational attraction of the material of the star upon itself. This balance kept the star in stable form until the fusion fuel was exhausted. Further changes in relative abundance, particularly of the heavier elements, occur in the cores and atmospheres of stars during outbursts.

A graph exhibiting the data points on relative abundances as the authors studied them in 1953 is given in figure 3.1. (The superimposed curves represent best fits to the abundance data for several starting conditions in the neutron-capture model of nucleosynthesis, which started with the Alpher-Bethe-Gamow paper in 1948 and was elaborated by Alpher and Herman in several papers through 1953.) Except in small detail, this data set continues to be useful. It has turned out that only light elements are produced in the early universe, including what had been a mysteriously high abundance of helium; heavier elements were produced and continue to be produced in stellar interiors and stellar outbursts.

George Gamow's contributions to this field began with an obscure paper in 1935. We return to Gamow later in this chapter.

Expansion of the Universe

The discovery of the expansion of the universe resulted from studies by Hubble, Humason, and Slipher working at the 100-inch Mount Wilson telescope. This observational discovery, ably pursued since Hubble's time by Allan Sandage, Gustav Tammann, and others, is surely one of the if not the major event in observational cosmology during the last century. The observations became possible with the advent of telescopes whose great improvement in light-gathering power made it possible not only to see faint, distant galaxies but also to look at the spectral composition of the light emitted by these objects.

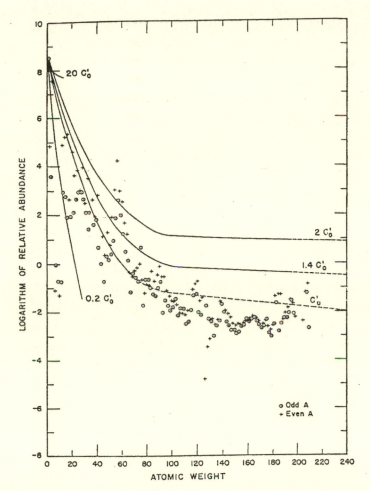

Figure 3.1. Comparison of observed cosmic abundances relative to the abundance of the element silicon, versus atomic weight, with results of simple prestellar neutron-capture sequence calculations in a standard Big Bang model. Starting conditions are $T \cong 0.11$ Mev (about 10^9 kelvin) and time $\cong 140$ seconds, with the variable parameter being the starting baryon concentration (in units of about 10^{17} cm^{-3}). (From *Annual Review of Nuclear Science* 2, © 1953 by Annual Reviews, Inc. Used by permission.)

Two kinds of measurement must be combined in order to observe the universal expansion. First, one must be able to estimate the distance of the object; second, one must be able to determine the spectrum of the light from the object: that is, analyze the wavelength content of the light and find how much the spectral features are shifted with respect to the wavelength of these same features as studied in the laboratory. Doing so in turn enables estimation of the velocity of the light source with respect to terrestrial observers if one attributes the spectral shifts to a Doppler effect. Again, as already mentioned, this effect is not a true velocity shift in spectral lines but, according to relativistic modeling, is to be interpreted as the expansion of the space-time in which the cosmic objects are embedded.

A word may be in order for readers who are not familiar with spectroscopy. Light is emitted by all systems above a temperature of absolute zero. Whatever the emitter may be, it will emit light at wavelengths that are characteristic of the elemental composition of its material. If the system is illuminated from behind by another emitter, then the system will absorb light at wavelengths characteristic of both that emitter and the material of the absorber. If the emitter is very hot and dense, then the emitted light will be a continuum, not exhibiting line structure. If we then look at the light through a dispersive medium (a prism or a grating made of fine, accurately spaced lines ruled on a reflective or transmitting material), and if, further, we place an opaque material as a blocking screen with only a narrow slit for the light to pass through, then we can observe images of the slit, spread out in wavelength, and we can identify the slit images at various wavelengths with the wavelengths of the particular elements.

There is much information to be obtained from these slit images. We can identify specific elements as absorbers or emitters, and from the intensity of light in the slit image we can in many cases infer how many emitting or absorbing elements are involved in producing the image. If we know what the spectral lines look like in a laboratory arrangement, and find that pattern in the spectrum we are observing, then we can infer the velocity of the emitter or absorber with respect to the observer. One final comment. If the source is dense, such as a dense gas or an emitting solid, the spectrum we observe may appear to be continuous (i.e., with no line structure). This may result from a thickening of the slit images due to the pressure, so that they overlap and the intensity appears continuously distributed in wavelength; or there may actually be continuous emission due to processes at high pressure and temperature other than those that form atomic spectral lines. For example, in low-pressure lamps, where emission results from an electrical discharge, we will see lines characteristic of the gases carrying the discharge, whereas in a high-pressure lamp we will see a continuous spectrum in the range of visible wavelengths, without line structure. In astronomy, one may see a background continuous spectrum from the emitter, but with absorption lines superimposed and created by intervening low-pressure systems. By way of illustration, a photograph of a spectrum from a low-pressure electrical discharge lamp filled with gaseous helium is shown in figure 3.2.

To return to the thread of this section: astronomers refer to a *cosmological distance ladder* in which the lower few rungs are reasonably well determined but the higher rungs involve less firmly founded extrapolations. There is a considerable element of "bootstrapping" involved in these extrapolations, but they are all that astronomers have at hand. Fortunately, there are frequently several different ways of establishing a given rung, which lends confidence in the observations. The reader should remember that astronomers are basically extrapolating their distance determinations from observations made in the solar system and the local galaxy in which it is embedded, the Milky Way. The range of distances to be covered is from hundreds and thousands of light years in the Milky Way to billions of light years for the faintest, most distant galaxies.

The first step in cosmological distance studies is to determine the distance from Earth to nearby stars by a parallax method, a straightforward application of trigonometry. Nearby stars are those that show a change in apparent position when seen from opposite ends of Earth's orbit around the Sun. Knowing the diameter of Earth's orbit from a variety of studies, we can use this diameter together with a measurement of the angular separation of the two sight lines to the star from the opposite ends of the orbit to triangulate for the star distance, a procedure common in surveying. It is fortunate indeed that some of the class of stars known as Cepheid variables are found at distances that overlap the region in which

Figure 3.2. Emission line spectrum of helium (the helium gas was in a sealed tube and excited by an electrical discharge). The broad line at the left is at 3,890 Angstrom units ($\times$ 10^{-8} centimeters); the faint line at the right is at 7,065 Angstrom units. Helium, a major constituent of the cosmos, was first discovered in the Sun. (The spectrum was taken in 1974 by an old friend, the late A. T. Goble, of Union College, using a glass prism in a Gaertner spectrograph.)

parallax can be measured. As first published by Henrietta Leavitt in 1912, a study of these variables found a regular relationship between the amplitude and the period of repetition of the fluctuation in their brightness. The amplitude actually seen is relative brightness and can be corrected to absolute brightness by means of the inverse square law, which states that the amplitude of light from a given source varies inversely as the square of the distance from that source. Now, given the period-luminosity relation from Cepheid variables, astronomers can recognize Cepheid variables by the periodicity of their light output, measure the apparent peak brightness and the time dependence of the brightness variation, and use the inverse square law to derive the absolute brightness and thence the actual distance. With the great ability of the Hubble Space Telescope to let us see very faint objects, Cepheid variables have been observed in galaxies many millions of light years away, overlapping other rungs in the cosmological distance ladder. It is appropriate to recall that Hubble's other great contribution to observational cosmology was his first observation of Cepheid variables in the galaxy called the Large Magellanic Cloud, whereby he established that certain faint objects thought by many astronomers to be clouds in the Milky Way are in fact galaxies like the Milky Way but enormously far off (about 170,000 light years for the Large Magellanic Cloud).

Other techniques being used to measure distance include, for example, a method based on the simple assumption that the brightest galaxy in any cluster of galaxies is equal in brightness to the brightest galaxy in a nearby cluster, such as the Local Group in which the Milky Way is embedded. Using the inverse square law again, we can deduce the distance of the brightest galaxy being studied. Another example derives from the observation that there appears to be a relation between a spiral galaxy's rate of rotation about its center of mass and its absolute brightness; this is the so-called Tully-Fisher effect. There are other approaches, such as using the Rashid Sunyaev and Yakov Z'eldovich effect, which we expand upon in chapter 5, and the agreement among the results of various approaches is getting better as time goes on. But even with reduced dispersion, the spread of values in such measures is the principal cause of uncertainty in determining the rate of expansion of the universe.

The other quantity that must be determined in the Hubble law is the velocity of galaxies or clusters of galaxies with respect to the observer. If there is enough light from a galaxy (for faint galaxies it may be necessary to make long exposures to get enough), the astronomer focuses the light through a narrow slit and examines its spectrum. Any shift in wavelength is attributable to the velocity of the source (more precisely, the integrated expansion of the expanding space matrix as seen along the line of sight to the emitting object). There is a fundamental relation

$$c = v \times \lambda$$

where the symbols are velocity c, frequency v, and wavelength λ, respectively. The shift in wavelength is related to velocity by

$$(\Delta\lambda)/\lambda = V/c$$

That is, the shift in wavelength as a fraction of the wavelength is equal to the velocity V as a fraction of the velocity of light in vacuum. Again it is a well-documented principle of relativity that c is numerically constant, regardless of the relative velocity of source and observer.

The initial observation of the radial velocity of a spiral galaxy was made by Slipher in 1912. By 1914 he had measured radial velocities for 15 galaxies; for most of these he found that the shifts in wavelength were toward the red, whence the name *redshift*, indicating recession. There were a few exceptions: the random velocities of some nearby galaxies with respect to the center of mass of the cluster of galaxies of which they are members apparently exceed their velocities as part of the universal expansion. Working with Humason and with Slipher's results, in which distances were not measured, Hubble determined distances to a number of galaxies, using Cepheid variables and other early techniques. At about the same time, George W. Ritchey and Curtis first used novae (stars that flare up in brightness from time to time) seen in other galaxies to determine distance. Banking on what one might call a principle of the uniformity of nature, these astronomers supposed that all novae achieved the same absolute brightness in their outbursts; they thus deduced distances by measuring the apparent brightness of such novae and invoking the inverse square law. From all these studies, Hubble was able by 1929 to propose a linear relationship between radial velocity and distance, leading to what became known as Hubble's law. It is interesting, historically, that Hubble based his synthesis on a sample of galaxies that were nearby and few in number. We should mention that models of the expansion of the universe suggest that at sufficiently large distances one might expect to observe some acceleration or deceleration in the expansion (i.e., a deviation from linearity), but the observational evidence for such a deviation is just beginning to come in. It suggests that there is an acceleration, consistent with the universe being open, and with there being a driving force due to a vacuum energy density related to the cosmological constant. We will say something more about this in chapter 6, which deals with the inflationary paradigm for the early universe.

The linear velocity-distance relation has been checked at a variety of wavelengths, including observations at radio-frequency wavelengths. The range of current observation vastly exceeds the range used by Hubble in formulating his relation. There is a discordant note in the interpretations of redshift: Halton C. Arp observes objects whose lateral separation projected in the sky is sufficiently small to suggest that these objects must be at about the same distance, yet he finds that they have disparate redshifts, whence he claims that the usual interpretation of redshift is not correct. Most astronomers, however, consider the number of such close associations of projected positions in the sky to be statistical artifacts, and the usual association of redshift with recession velocity is widely accepted as the only physical explanation of the phenomenon.

The linear Hubble relation is

$$V = H \times D$$

where astronomers usually express the velocity V in kilometers per second, and the distance D in megaparsecs. A megaparsec is a million parsecs, the parsec being a distance unit equal to 3.26 light years. A parsec is the distance of an object that has a proper motion across the sky which subtends an angle of one second of arc as seen from opposite ends of Earth's orbit. The quantity H, variously called the Hubble parameter (which it is) or the Hubble constant (a constant locally but not cosmologically), is the rate of expansion of the cosmos, and has units of kilometers per second per megaparsec (one kilometer per second per megaparsec equals 3.24×10^{-20} sec^{-1}). As measurements of distance improved over the years, the Hubble parameter dropped from the enormously high value of over 600, first proposed by Hubble, to a value between 50 and 100. This range persisted for a number of years but has now been reduced to about 50 to 75 kilometers per second per megaparsec. Hubble's very large initial value was corrected in part in about 1952 when Walter

Baade at Mount Wilson corrected an error in Hubble's use of Cepheid variables in distance calibrations.

The list of publications of measurements of the Hubble parameter is pages long. At the time of this writing, teams of astronomers are working diligently on establishing the Hubble parameter value using measurements made with the Hubble Space Telescope. Establishment of an improved distance scale continues to be one of the major stated missions of the instrument. For example, two teams working with HST from the California Institute of Technology anticipate that the discrepancy between their measurements will be significantly lessened within the next few years or less, particularly now that HST has been recently serviced. One team, led by Wendy Freedman, has proposed values for the expansion rate of the universe that until recently have been high enough to cause concern because the derived age of objects in the universe may be greater than the Hubble age. Another team, headed by Allan Sandage, finds values for the expansion rate that are compatible with a reasonable age of the universe. In the last few months (early in 2000) Freedman's group have papers in press giving Hubble parameter values (all in units of kilometers per second per megaparsec) of 70, 71, 71, 69, and 68, and have established a calibration for the Tully-Fisher approach, which yields 68, as does yet another approach using surface brightness fluctuations. To give a feeling for the effort involved, consider that the number 70 came from data on 18 galaxies and 800 Cepheid variables. Sandage's team consistently reports values in the range 50 to 55. Both teams are trying to ensure the calibration of the Cepheid period-luminosity relation which they use by checking with other distance-measuring techniques.

For example, one technique of great interest and promise is the observation that a particular class of supernovae (exploding stars, readily seen at great distances) all appear to achieve about the same absolute brightness during the explosion process, another invocation of the principle of the uniformity of nature. A group of astronomers at the Smithsonian Astrophysical Observatory, led by Robert Kirshner, is using this approach and has recently reported a value of 64. Finally among the techniques we should mention is one using the effect named after the two scientists who poposed it, namely, Rashid Sunyaev and Yakov Z'eldovich. The cosmos contains everywhere the ubiquitous CMBR at 2.73 kelvin. The photons of this radiation interact with hot electrons in space inside clusters of galaxies and are scattered to higher energies. Radio observations of the global CMBR, which include galactic clusters, will exhibit a hot spot, known as the Sunyaev-Zeldovich effect. One can derive from the image the size of the cluster, its included angle of view, and, thence, the distance. The red shift of the cluster's light yields a velocity and a derived Hubble parameter. It is a work in progress, but the results are consistent with those of the other teams. Yet another group has recently reexamined the question of the age of galactic clusters and found, on average, ages of about 14 billion years, within a range of 12–16 billion years.

There are errors associated with the Hubble parameter observations, some random and some systematic. There is no point in a detailed discussion of these errors here since they do not affect the message we are trying to convey; rather, suffice it to say the errors show overlap in the various Hubble parameter measurements.

It is interesting, though, to plot the values of the Hubble parameter against the year of observation, as shown in figure 3.3. A convergence of all modes of observation definitely appears to be in the cards.

One recent study of the distance of a galaxy has been carried out by straightforward triangulation, without dependence on Cepheid variables as a yardstick. The results suggest that distances based on Cepheids may have to be revised, perhaps by as much as 10%. Moreover, triangulations done by the HIPPARCOS satellite have also suggested a need for reexamining Cepheid variable distance scales.

A final remark about the Hubble law: as is so often the case in science, work by others before and during the period when the Hubble law was being established came very close to establishing a velocity-distance relation. In particular, there was work by Carl Wirtz and K. Lundmark, which is described in the book *Modern Cosmology in Retrospect* (1990) edited by B. Bertotti et al. (see the recommended reading list). Neither, however, made the great leap in synthesis of results which was the contribution of Hubble.

The calculation of the age of the universe according to the Big Bang model requires a knowledge of the value not only of the Hubble parameter but also of the mean density, Ω, of matter in the universe. Depending on that value, there are three possibilities for the age in the standard Big Bang model (the equations for age are given in the appendix). We can define a critical density for the Big Bang model as one that would lead eventually (after an indefinitely long time—infinite, in mathematical terms) to a halt in the expansion. If the mean cosmic density is less than

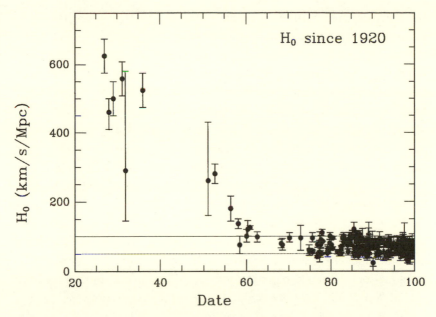

Figure 3.3. The present value of the Hubble parameter, H_0 (the expansion rate of the universe in units of kilometers per second per million parsecs, a parsec being 3.24 light years), plotted against the year of measurement. It has been the subject of many measurements over the years which have been difficult to make, but with technological improvements in recent years the values are converging to the range 50 to 75. The Hubble parameter, together with measurements of the mean smeared-out density of matter in the universe, are important cosmological parameters that will help to tie down the nature of the cosmological model. (Supplied by John F. Huchra of the Center for Astrophysics, Harvard University, and used with his permission.)

critical, the expansion continues indefinitely; if it is more than critical, the model predicts that the expansion will stop and the universe then go into a state of contraction, which would lead ultimately to what is sometimes called the Big Crunch. The model also admits oscillation: that is, cycles of successive Big Bangs followed by Big Crunches.

The last word is not yet in, either on the value of the Hubble parameter (we come down on the side of Sandage and his collaborators, who propose a Hubble parameter of about 55 kilometers per second per megaparsec) or on the mean density of matter in the universe. Most astronomers use the range of values of these two observational parameters in stating estimates of the age of the Big Bang model. If the reader sees an age quoted on the basis of a measurement of the Hubble parameter, without specification of the mean density, the writer is either selling the reader short or is implicitly assuming the reader is understanding that the result refers to an $\Omega = 1$ model of the universe. Many astronomers and science writers do not even specify what value of the Hubble parameter is being used; at best they give the redshift observed, which implies an age, or simply quote the age result. We submit that this kind of approach does an injustice to the cosmological literature, technical or otherwise. Some astronomers and many science journalists assume that the readers of the reported results will understand that they prefer a particular value of the density of matter; in particular, they argue for a ratio of unity for the total density compared with the critical density: that is, a density which specifies that the expansion behavior of the universe should lie exactly midway between an open model and a closed model. Note that in the standard Big Bang model the density ratio Ω is for all practical purposes is unity until matter and radiation decouple in the expansion, at a time of the order of 300,000 to 500,000 years (see the appendix). Some proponents of an inflationary model of the early universe require that the universe be at $\Omega = 1$ for all time, at least for philosophical or aesthetic reasons. Early on, the argument was made that since the observed properties of the universe require that Ω have a value of something less than unity but not greater than 2, say, in order that the rate of formation of objects in the cosmos be neither too fast nor too slow compared to the estimated age of the universe, then an exact $\Omega = 1$ model might be reasonable.

The appendix presents a graph that enables one to read off the age of the universe in a Big Bang model, given specified values of the mean matter density and the Hubble parameter. We will say a bit more about discrepancies in measurements later in this chapter when we discuss the question of the age of the universe in connection with our studies of the Big Bang model in the late 1940s and early 1950s. As already mentioned, we had a period of considerable concern about this model because the then current value of the Hubble parameter, used together with an estimate of the mean density of matter in the universe (also due to Hubble), led to a time since the Big Bang which was less than the much better known age of the Earth. Many cosmologists dismissed the Big Bang model because of this discrepancy, although with hindsight one might argue that predicting an age of some billions of years in the late 1940s and early 1950s was already far better than previous models had managed.

In any event, the discrepancy was a major deterrent to widespread acceptance of the Big Bang model at the time, and led to the ascendancy of the Steady State model for some years. It was the mid-1950s before the cosmological parameters involved in age calculations began to admit of more reasonable age values. There was a particularly difficult period when a German astronomer named Albert Behr published cosmological parameters that really confused the issue, particularly when we used his parameters in our method of estimating the temperature of the microwave background radiation (which we say more about in chapter 5), but those values were superseded by other measurements within a relatively short time. Criticism of the Big Bang model by the Steady State school was in our view unduly harsh, given the difficulties inherent in determining the requisite parameters. (The Steady State model and other alternatives to the Big Bang are discussed in chapter 4.)

Development of Mathematical Nonstatic Models

While Einstein's cosmological model was current, Willem de Sitter in the Netherlands published several papers in which he distinguished between, on the one hand, models of the universe that were finite but unbounded (as in Einstein's basic model, which resembles a sphere in three-space or a four-dimensional cylinder in space-time) and, on the other hand, an infinite, unbounded model. He found that he could model a universe that was essentially devoid of matter but with a nonzero cosmological constant. In this model the proper distance (i.e., the separation of any arbitrarily selected pair of points in space-time) depended on the value of the cosmological constant. Such an empty model was found to have some interesting properties, exhibiting, for example, a temporal redshift of any radiation injected into the otherwise empty model. The work of Wirtz and Lundmark, already mentioned, as well as the early work of Hubble, assumed the de Sitter model as a working model for understanding the relation of velocity and redshift, and investigators have continued to deal with it as an approximation to the real universe with its very low mean matter density. In retrospect, it seems remarkable that the de Sitter model foretold the possibility of a cosmological model exhibiting expansion of the universe before expansion was an accepted feature.

Some years after the Einstein model was described in the literature, Alexander Friedmann, in the Soviet Union, reexamined the Einstein model using an arbitrary cosmological constant whose values could be positive, negative, or zero. He developed models with inherent expansion or contraction, their precise behavior dependent both on the mean density of matter in the universe and on the value assigned to the cosmological constant. (Friedmann's work published in 1922 dealt with positive spatial curvature; in 1924, with negative spatial curvature.) He found that he could recover the Einstein static model, the empty de Sitter model, and a large number of other models as well, depending on the cosmological constant. Nonstatic solutions were also explored by Georges Lemaître, a Belgian cleric, in 1929.

A curious historical footnote is that the first paper by Lemaître does not mention the work of Friedmann in its text, although it does list the first of his two seminal papers in a bibliography. The conventional wisdom is that nonstatic solutions were indeed developed independently by Friedmann and later by Lemaître. Perhaps the Friedmann reference was inserted by a reviewer or editor, perhaps Eddington, and Lemaître had not in fact seen the Friedmann work, since the mathematical expressions for the rate of expansion of the universe given in the papers are in fact the same. However, we shall continue to refer to the Friedmann-Lemaître equation as the basic equation of the Big Bang model, leaving the resolution of this confusion, if in fact there is any, to some future historian of science.

As the idea that the universe was expanding came to be widely accepted, with Hubble's seminal paper defining Hubble's law appearing in 1929, Lemaître was the first to discuss some of the physics involved in a nonstatic model. He tried to reconcile the model with Hubble's observed expansion rate and mean density of matter in the universe. He recognized that the Einstein model with a cosmological constant admitted of an arbitrarily large amount of matter that was static, whereas the de Sitter model was devoid of matter but admitted of expansion or contraction—that is, was nonstatic.

The bottom line in Lemaître's work at the time was his particular interest in explaining the origin of cosmic rays, high-energy particles whose cosmic origin was recognized—as the result of pioneering observations by Victor Hess of Fordham University— but not understood. The proposal made by Lemaître was that the cosmos began as an all-encompassing, gigantic primeval nucleus (in French, the "primeval egg") which broke up into nuclear-sized pieces. Some of these nuclear pieces came away from the breakup with very high energy, he suggested, and survived to be identifiable as cosmic rays. Current views are that cosmic rays originate in supernova explosions (ultra-high-energy cosmic rays are now considered as possibly originating in gamma-ray bursters, whose nature is still very mysterious at the time of this writing). Lemaître's largely qualitative model was not widely accepted. Many years later (1948) Maria Göppert-Mayer and Edward Teller, both then at the University of Chicago, again proposed a single cosmic nucleus whose breakup served as the origin of all matter and energy in the universe. In any event, it appears that Lemaître should be credited as the first to try to introduce some physics into the connection between the beginnings of the universe and present-day observations. It also appears that the next scientist to try to incorporate physics into cosmological modeling was George Gamow.

George Gamow's Early Foray into Cosmology

As already mentioned, George Gamow, who was to become a most valued collaborator with us on cosmological problems, in 1935 published a paper in the *Ohio Journal of Science*, a somewhat obscure journal, in which he noted with interest the then-recent discovery of the neutron and the consequent study of neutron-capture reactions done for a variety of nuclear species by Enrico Fermi, then in Italy. This work led Gamow to suggest the importance of neutron-capture reactions in establishing the abundance distribution of nuclear species, and the undoubted role of neutrons in reactions producing energy in stars. In 1942 he gave a lecture (briefly abstracted in the *Journal of the Washington Academy of Sciences*), again suggesting that the elements were formed somehow in nuclear reactions in a system not in thermodynamic equilibrium. In 1946 Gamow published a more specific set of ideas on nuclei being formed by successive neutron-capture reactions and chose as the locale of formation the early universe. (Alpher was already at work on a dissertation with Gamow at that time.) He was motivated toward this end by the failure of equilibrium theories of element synthesis.

Equilibrium theories of element formation received a great deal of attention in the 1930s and early 1940s as researchers attempted to use the inherent stability properties of nuclei to understand their abundance. As we have already mentioned, every species of nucleus except hydrogen has a mass defect: that is, a difference in mass between the nucleus as a whole and the sum of the masses of contained neutrons and protons. Suppose a locale populated with a mix of neutrons, protons, and electrons at a temperature and density high enough so that the mass defect would drive the mix of elementary particles toward nuclear species, with the greatest amount of available energy bound up in nuclei as the energy equivalent of the binding energy; that is, the system would move toward an equilibrium state. Then the resulting abundances of nuclei should reflect their inherent stability. Higher binding energies signified greater stability. Calculations based on this premise did reasonably well in explaining the relative abundance of isotopes of a given species of nuclei, isotopes being nuclei having the same number of protons but different numbers of neutrons. Moreover, such calculations did reasonably well in dealing with nuclei near each other in atomic weight (sum of neutrons and protons, less the binding energy). It did not prove possible, however, to find a locale—defined by a single temperature and density of matter—which would make it possible to represent the entire relative abundance distribution of nuclei as shown in figure 3.1. If one found a locale suitable for the light elements, then calculations of the abundance of the heavier elements showed a great deficiency, a situation that Gamow labeled "the heavy element catastrophe," which is illustrated in figure 3.4. The last and most definitive paper on equilibrium theories was by Subramanyan Chandrasekhar and Louis Henrich in 1942, who suggested that it would be necessary to look at some kind of nonequilibrium processes. Moreover, if the locale were to be stellar interiors, it was not at all clear how one would get the abundance distribution to survive from the interior into interstellar space without serious modification. Nevertheless, it is now widely accepted, on the basis of much evidence, that the heavier elements are made in stars.

In 1946 it was still Gamow's hope to find a single locale for explaining the entire abundance distribution. Given the difficulties with single-locale equilibrium theories, he posited that the early expanding universe would have physical conditions suitable for nuclear reactions to occur in a nonequilibrium fashion (not unlike a chain reaction in a thermonuclear bomb) over a short period of time and then be quenched as the starting material was used up and diluted by the universal expansion. He went on to propose that nuclei would be built up from an initial neutron gas by some means of agglomeration, (the choice of a starting gas of neutrons only was parsimonious) together with neutron decay into protons and electrons, the final nuclear states being arrived at by intervening conversion of neutron-rich fragments into stable nuclei by radioactive beta decay: that is, by the emission of electrons from the fragments.

In 1945, Alpher had completed a master's thesis on sources of energy in stars, his first in-depth consideration of an astrophysical topic. His

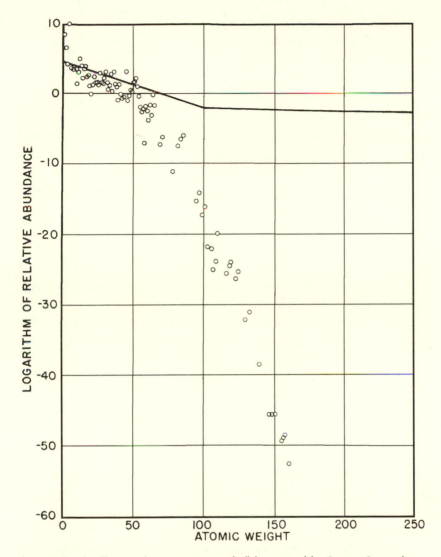

Figure 3.4. The "heavy-element catastrophe" (so named by George Gamow). Using simple equilibrium theories to explain observed relative abundances—theories which require specification of single temperature and density and which are based on the binding energies of atomic nuclei (see figure 7.1)—one finds that a representation for the lighter elements leads to an enormous deficit in the abundances of heavier elements. Note the logarithmic scale. This seems to rule out simple equilibrium theories to represent cosmic abundances. (From *Reviews of Modern Physics* [April 1950]: 153–212, used by permission of the American Physical Society.)

mentor was Gamow, who then accepted him to work on a Ph.D. dissertation. The first mutually acceptable topic was the development of structure in the universe, requiring calculations of the behavior of various kinds of density perturbations in a relativistic, homogeneous, isotropic expanding universe containing matter only. The work drew heavily on the general theory of relativity; Alpher carried it on with some success, arriving at the conclusion that small perturbations of the density could grow but not on any cosmologically useful time scale. In late 1946, unfortunately for us, Gamow received a then current issue of the *Journal of Theoretical Physics of the USSR* containing a paper (also a Ph.D. dissertation) by Evgeny Lifshitz on the same subject and with the same conclusions. Alpher has a vivid recollection of Gamow coming into his office, waving a copy of the journal, and saying, "Ralph, you have been scooped." In one of a number of silly things Alpher has done, he destroyed his voluminous notes on this subject (with pages and pages of perturbed Riemann-Christoffel brackets) and decided, with Gamow, on a second dissertation topic: developing Gamow's rather cursory 1946 ideas on primordial nucleosynthesis in the early stages of an expanding universe. Incidentally, Lifshitz went on to a distinguished career in theoretical physics and coauthored a superb series of books with the well-known Soviet physicist Lev Landau, who had coauthored with Gamow a paper—published during the year Gamow came to the United States, 1933—on the temperature in stellar interiors. It's a small world.

It may help the reader to some perspective on these matters to recount where Alpher, Gamow, and Herman had come from and where they were when the Big Bang model work began. George Gamow had come to George Washington University in Washington, D.C., in 1933 from the then Soviet Union; he was already deservedly known internationally for his papers explaining alpha radioactivity and the theory of nuclear reactions involving charged particles (work for which he should have received the Nobel Prize, or at least shared with E. U. Condon and G. Shortley, who independently published on alpha decay a week after Gamow did). No Nobel Prize was awarded for the theory of alpha decay. While at George Washington University he served as a consultant to the U.S. Navy Bureau of Ordnance during World War II and after the war at the Applied Physics Laboratory of Johns Hopkins University (APL/JHU). He left George Washington in 1956 to join the physics faculty at the University of Colorado in Boulder.

Gamow was unable to participate during World War II in the development of nuclear weapons because of an absurd question of security: in the Soviet Union he had become a commissioned officer in the Red Army, which made it possible for him to serve as a professor of meteorology at the Soviet equivalent of West Point. Those who knew Gamow also knew that he had nothing but scorn for the Stalinist regime, which removed his name from the Soviet Academy of Sciences and sentenced him to death in absentia for leaving the USSR. He did finally receive clearance after the war to participate in research at the nuclear weapons laboratories at Los Alamos. And he was reinstated as a member of the Russian Academy of Sciences posthumously, probably at the instance of the well-known Russian physicist and cosmologist Yakov Z'eldovich.

Gamow's work for the U.S. Navy (which did grant him security clearance but not for the Manhattan project) was concerned with the optimization of the packing of explosives, work in which he collaborated with the theorist John von Neumann, later best known as the father of the stored-program digital computer and for his definitive studies of the foundations of quantum mechanics. Among other matters at the APL/JHU, Gamow and Alpher, with a third researcher named James Pickering, developed a model for a ramjet missile using radioactive waste material as a heat source. When the work was submitted for publication, the navy and the FBI appeared on the scene, told us our work was classified beyond our classification status, and confiscated our notes. Gamow had some remarks to make which were, we are sure, choice, but in Russian. At that point, Alpher resolved that in his next job, if any, he wanted nothing to do with the military and classification. Unfortunately, things did not work out that way.

Alpher worked at the U.S. Naval Ordnance Laboratory and at the Navy Bureau of Ordnance from 1940 to 1944, where among other things he served in a group headed by John Bardeen, who later became quite famous (later twice a Nobel laureate for his work on the transistor and on superconductivity theory). In 1944, Alpher moved to the APL/JHU. During this period he finished a bachelor's degree as a night student at George Washington University and got to know Gamow both in the courses he taught and at the Navy Bureau of Ordnance. After 1943, Alpher continued night graduate courses at George Washington and from 1943 to 1945, working with Gamow, wrote the master's thesis on energy production in stars; it was submitted on V-E Day, 1945. For the navy, Alpher had worked on the protection of ships against magnetic mines and on magnetic airborne detection of submarines. When he moved to the APL/JHU in 1944, his work involved developing and getting into production a magnetic gradiometer proximity exploder for air-launched torpedoes. In 1945 he stayed on there in a program to develop ground-launched antiaircraft guided missiles. His particular contributions were initially in supersonic aerodynamics. Meanwhile, his contacts with Gamow continued, as both student and colleague; Alpher received his Ph.D. from George Washington University in 1948, while he was at the same time a full-time employee of the APL/JHU.

Herman, a graduate of the City College of New York, where he worked extensively with Mark Zemansky and Henry Semat, received his Ph.D. in physics from Princeton in 1940, using infrared spectroscopy to unravel the structure of fatty acid molecules. (Alpher notes that Herman gave up a promising career in art when he turned down a fellowship at the Pratt Institute; moreover, he participated in some very creditable work with a young psychiatrist named Milton Sapirstein on the effect of drugs on the central nervous system.) He began his doctoral work with Edward U. Condon, but Condon left Princeton before Herman had completed his dissertation, and Herman essentially did the whole thing on his own. While in graduate school he took a course in relativity and cosmology with Howard P. Robertson, who was by then a recognized authority in the field. Robertson also chaired Herman's dissertation defense committee. Herman then spent a year at the Moore School of Electrical Engineering of the University of Pennsylvania, working on the early problems of a digital computer that was ultimately built there. In 1942 he joined Section T of the Office of Scientific Research and Development, a wartime agency that undertook to develop proximity fuses for antiaircraft shells and rockets. The work was done initially at the Department of Terrestrial Magnetism, Carnegie Institution of Washington, but in 1943 the activity was split between nonrotating vehicles (rockets), at the National Bureau of Standards, and rotating projectiles (artillery shells), which became the main raison d'être of the APL/JHU. Herman's chief concern there during the War was quality control on the performance of the proximity fuses being developed.

Alpher met Herman shortly after arriving in 1944. When the war ended, they both joined a newly formed research center, with Herman doing mostly infrared spectroscopy and combustion kinetics, and Alpher working in supersonic gas dynamics and later studying primary cosmic radiation in a group headed by James A. Van Allen, now well known for his discovery of trapped particles near Earth, the so-called Van Allen belts. When we were planning to leave the APL/JHU in 1955, we had a joint interview at the State University of Iowa, where Van Allen had become head of the physics department. For a variety of reasons, Herman and Alpher declined positions at Iowa and went elsewhere, separately. In the late 1940s, however, there was a commonality of interest in astrophysics and cosmology, and as Alpher developed his several dissertation topics, discussions ensued with Herman and, of course, with Gamow. (Alpher notes that a number of his contacts with Gamow were at a bar and grill called Little Vienna, on Pennsylvania Avenue near George Washington University, usually for an hour or so before Gamow went off to lecture. The cafe is gone, but the memory of our meetings, and of the libations prior to the lectures, lingers on.)

The Alpher-Bethe-Gamow Paper

Given the appearance of Lifshitz's work, the Alpher-Bethe-Gamow paper was thus Alpher's second foray into a topic related to cosmology. What Gamow and Alpher chose to do was to see if one might understand the synthesis of the chemical elements in a hot, dense early phase of the expanding universe. The success of such a program would be judged by how well the calculations matched the observed cosmic abundances of the elements, where such abundances are considered as having been established in the early universe and as therefore constituting part of the remains of early cosmic history.

Our mathematical model for consideration of the early universe was based on the theoretical studies of Friedmann and Lemaître of a nonstatic model. Given the expectation of high ambient temperatures early on, we anticipated that the total density of the universe would have been approximately the density of the radiation present and, in fact, that the dynamic behavior of the universe at early times would have been governed by the radiation density. It was not clear to us initially how much of the total density to ascribe to the matter present, but we thought it would make a very small contribution, with its numerical value determined by the conditions for synthesizing the elements. The density of radiation, as usually stated in mass per unit volume, was millions of times the density of matter present at, say, one second of time into the expansion. The main obstacle initially was our lack of detailed knowledge of

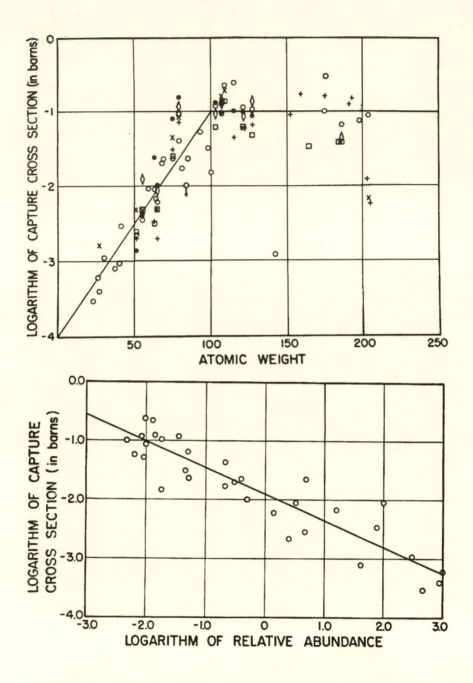

Figure 3.5. Top figure: measurements of the probability of so-called (n, γ) reactions as published by D. Hughes, then at Argonne National Laboratory. These are reactions in which a neutron is absorbed by a nucleus and excess energy is emitted as a photon; they are called neutron-capture reactions. The measurements were made at ≈ 1 MeV; Hughes was surveying all manner of

the probability of various kinds of nuclear reactions among the species likely to be present at the onset of nucleosynthesis.

We had just emerged from World War II, and data for such reactions—reactions driven by high temperatures and densities—were still classified or just beginning to be declassified from the Manhattan Project. As luck would have it, Alpher and Gamow got a great boost from work published by Donald J. Hughes, then at Argonne National Laboratory and later at Brookhaven National Laboratory. Hughes had set out to survey any and all materials that might be of interest in reactors: controlled chain-reacting ensembles of fissionable materials in which large fluxes of neutrons are produced. These neutrons are effectively at temperatures of 1 MeV (million electron volts), or about 10^{10} kelvin, which incidentally would have been the cosmic temperature at about a second into the expansion. Hughes bombarded targets of a large number of materials with 1 MeV neutrons. What he measured was the probability that a nucleus would absorb a neutron and become a new nucleus with an increased atomic weight (due to the added neutron) of unity but with the same atomic number (content of protons). The new nuclei rid themselves of excess energy by emitting gamma radiation. These so-called neutron-capture reactions cause modifications and perhaps degradation in materials in and around reactors. The exciting result of Hughes's work is shown symbolically in figure 3.5. (Please refer back to figure 3.1, which gives the course of the cosmic relative abundances.) The top diagram shows the variation with atomic weight of the probability of neutron-capture reactions, where we can see the exponential rise of the probability to an atomic weight of about 100, and essential constancy for heavier nuclei, mirroring the properties shown in figure 3.1. The second diagram shows a correlation of the probability of neutron-capture reactions with abundances. Designation of reaction cross sections in barns (10^{-24} cm^2) was a World War II innovation.

On the basis of Hughes's suggestive work, Alpher and Gamow calculated in a very approximate way the growth of abundances by neutron-

nuclei to identify those suitable to be used in a nuclear reactor environment. (The plot also includes other data available in 1950.) The bottom figure is a correlation of the reaction cross sections with the observed cosmic relative abundances of the particular nuclei. A barn is 10^{-24} cm^2. (Reprinted by permission of the American Physical Society from *Reviews of Modern Physics* 22 [April 1950]: 153–212.)

capture reactions. For simplicity, we assumed that the initial material was a neutron gas and chose a density of matter to give the best subsequent representation of observed abundances. We further assumed that the probability of neutron-capture reactions mirrored the general run of the probabilities shown as the second plot of figure 3.5, with the successive reactions beginning with the capture of a primeval neutron by a newly formed proton resulting from the radioactive decay of neutrons (free neutrons are radioactive, with a half-life of about 12.8 minutes). The nucleus thus formed would have been a deuteron, one of the heavy isotopes of hydrogen; in a simpleminded view, the deuteron would then have absorbed a neutron to form a triton, the nucleus of tritium, the heaviest isotope of hydrogen. Further neutron captures would have occurred. When nuclei so formed had an overabundance of neutrons and were therefore unstable to radioactive decay, intervening beta decay (the emission of an electron by the nucleus) would have adjusted the relative number of neutrons and protons to one neutron fewer and one proton more in the nucleus. The successively heavier nuclei that formed would in turn adjust by beta decay so as not to be overly neutron rich and would be stable for times exceeding the duration of nucleosynthesis.

For purposes of a calculation that one could carry out at a time when digital computers were rarities, it was assumed that the reaction rates would be fast compared with the rate of expansion of the universe, and therefore the calculations dealt basically with a static model, a useful approximation at the time. This very simplified calculation led to what then seemed a very exciting representation of the overall cosmic abundances of nuclei, as illustrated in figure 3.1. Moreover, the calculation rationalized the high relative abundance of helium, which had been a puzzling aspect of the relative abundance data.

The Alpher-Bethe-Gamow paper was published on April 1, 1948, and, despite its shortcomings, was rather widely viewed as the beginning step toward a possible model of the universe; other researchers quickly renamed it the alpha-beta-gamma model. There are several amusing anecdotes associated with the work—notably, the fact that the famous physicist Hans Bethe was associated with it in name only. It was as a bit of a joke—and against Alpher's preference at the time—that Gamow insisted on adding Bethe's name (for euphony) with the notation "in absentia." The editor of *Physical Review*, to whom the manuscript had been sent, cleared the use of his name with Bethe, who later remarked that he had no objection for the reason that the work might even be right; moreover, "in absentia" was removed in the printed version.

One consequence was that Bethe was persuaded by Gamow to serve on Alpher's dissertation examination committee at George Washington University in the spring of 1948. This was a bit of good-hearted blackmail, for Gamow reminded Bethe of his involvement in the early 1930s with a famous paper that spoofed the work of the English astronomer Sir Arthur Eddington. That paper, concerning Eddington's ongoing efforts to develop numerical relationships between the absolute kelvin temperature scale (on which 0°C is 273 kelvin) and a famous constant of atomic physics, the fine-structure constant, was published in the letters section of *Die Naturwissenschaften*, a serious German science periodical, and Bethe was actually invited to give serious talks on the subject before it was recognized as a spoof. Later, the German magazine published a cautionary statement to the effect that the editors were not responsible for the content of the letters section, which did not then receive detailed peer review.

Another consequence was that Alpher's dissertation defense, a rather formal affair with Alpher, Gamow, and examiners marching in academic regalia, was undertaken before an audience of some 300 people. Dissertation defenses were routinely advertised in university circulars but seldom drew such an audience. Among those present was the late Watson Davis, then editor of *Science News Letter*, which has become *Science News*; he wrote a brief news item that was syndicated in the national press. Since Alpher's dissertation dealt with matters in the early universe and the origin of the abundance distribution of the elements, the news release elicited well-meaning responses from all kinds of people. There were comments by some political cartoonists, and offers of special prayers and novenas from religious fundamentalists who took exception to the cosmic model portrayed. Little did they foresee that in 1952 Pope Pius XII, speaking to a meeting of the Vatican Academy of Sciences, would endorse the Big Bang model, whereas shortly thereafter, Sir Harold Spencer-Jones, Astronomer Royal of Great Britain at the time and a member of the Church of England, would endorse the then competitive but now generally considered unacceptable Steady State model of the cosmos.

Despite the deficiencies in the alpha-beta-gamma paper, it is still cited as the first foray into a physics-based cosmological model and as a forerunner to later developments in the Big Bang model.

Standard Big Bang Models Studied by
Alpher and Herman

After the alpha-beta-gamma manuscript had been sent in, Alpher and Herman proceeded immediately to reconsider the entire calculation, intending to evaluate the effect of removing some of the assumptions that had been made. (This cosmological work was done in our spare time, as we were both occupied full time in other work at the APL/JHU.) Chief among the assumptions removed was that the model was static; the effect was to reduce the reaction rates as the universe expanded, diluted, and cooled. We continued to use the smoothed representation of reaction rates as a function of atomic weight, even though we were concerned, and rightly so, that the rates among the light elements were much more complicated than in our representation. In addition, there was a particular problem, recognized early on, that nature had not provided us with sufficiently stable nuclei at atomic weights 5 and 8, so the notion of a sequential buildup of nuclei by neutron capture beginning with neutrons and protons could not be correct in detail for the light elements.

The one parameter whose value we were free to choose in model calculations was the density of matter at a particular epoch in the expansion, and it was satisfying to find that the value we needed with the more nearly correct calculations was close to the value chosen in the alpha-beta-gamma paper. Incidentally, a colleague of ours in the physics department of Johns Hopkins University, the late Theodore Berlin, provided an approach to the solution of the equations governing the growth of the relative abundances, a result that we published as an addendum to one of our publications of the period.

In late 1949, Alpher gave a colloquium at which Enrico Fermi of the University of Chicago was in the audience. Fermi was famous for many things, not the least his direction of the construction and first successful operation of a nuclear reactor under the grandstand at the university in 1942, a feat that was fundamental for the Manhattan Project. Fermi was intrigued by our concern about the lack of nuclear reaction data for the light elements; when he returned to Chicago, he enlisted his colleague Anthony Turkevich, a nuclear chemist who had also been involved with the Manhattan Project, and between them they developed a list of some 28 reactions among the light elements. They used observations of reaction rates, where they were available, and nuclear theory and educated intuition to estimate other rates. We show the interesting results of their nucleosynthesis calculations in figure 3.6. Fermi and Turkevich also used a static model approximation, and chose temperatures and matter densities similar to those in the alpha-beta-gamma paper. They sent their results to us to be included in an extensive review paper we wrote in 1950 on the general problem of the origin of the elements. They too were stymied by the gap occasioned by the lack of nuclei at atomic weights 5 and 8, as was also Eugene Wigner, a famed nuclear physicist. In 1952, Gamow drew a cartoon showing Wigner jumping a crevasse in the atomic weight distribution at mass 5. (This cartoon is figure 17 in Gamow's 1952 book *The Creation of the Universe*; Gamow also included in his figure 15 a cartoon representation of Alpher and Herman, based on a drawing done by Herman for the magazine *Physics Today*, titled "Cooperative Research.")

We had begun our work using electrified Marchant and Friden desk calculators, with the help of a "computer" (as women of the period who worked on numerical calculations by hand or on desk calculators were dubbed) named Kaki Pace, now Anderson. We graduated to a Reeves analogue computer, then to a MADDIDA (a magnetic drum storage computer developed by North American Aviation), next to an IBM CPC programmable card punch calculator, and, finally, to the SEAC, the first digital computer delivered to the National Bureau of Standards for use by the Census Bureau in 1950. On the IBM machine, located at the Watson Laboratory near the Columbia University campus, we had a good deal of programming help from L. H. Thomas, a theorist well known for his work on the Thomas-Fermi model of the atom.

Starting Conditions for Nucleosynthesis

It was clear from the beginning in our calculations that starting the reaction processes for building up nuclei with a neutron gas, as described above, was another oversimplification. A major step forward came with the work of Chushiro Hayashi of Japan in 1950. The universe was surely hotter and denser before nucleosynthesis began; in all earlier calculations, including those of Fermi and Turkevich, the ratio of neutrons to protons at the start of element building was taken to be that which had resulted from the free decay of primordial neutrons prior to a specific starting time. Hayashi proposed instead that the ratio of neutrons and protons should be taken to be whatever resulted from spontaneous and induced beta-decay processes in the early stages of the expansion, in the presence of electrons, positrons (the antimatter equivalent of electrons), neu-

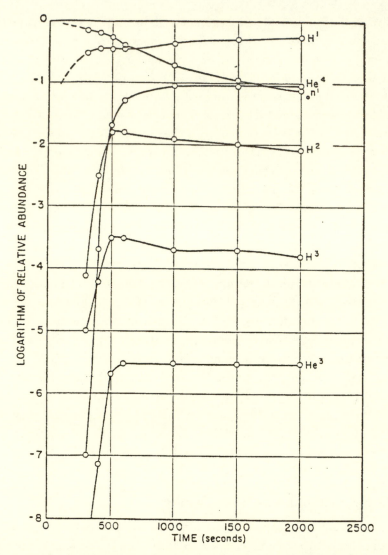

Figure 3.6. Calculations by Enrico Fermi and Anthony Turkevich, both then at the University of Chicago, of prestellar relative abundances of the light elements, using specific nuclear reactions either as measured in the laboratory or as calculated (they considered 28 reactions). Starting conditions were as in figure 3.1. (First published by Alpher and Herman in *Reviews of Modern Physics* 22 [April 1950]: 153–212, and reprinted by permission of the American Physical Society.)

trinos, and antineutrinos. Element building would then proceed with the consequent neutrons and protons present as the temperature dropped low enough to render unimportant the photodissociation of nuclei by collisions with the abundant supply of energetic photons present.

Unfortunately Hayashi's calculation gave values of the neutron-proton ratio that precluded a successful generation of the relative abundance distribution in our simple overall neutron-capture picture. To correct what we felt were difficulties with his calculation and to see whether an improved calculation would yield useful abundances, we worked through Hayashi's approach, using relativistic corrections for reaction rates at the high energies involved in the reactions, using a more nearly correct relativistic model (because the expansion rate was modified by the inter-conversion of matter and radiation at early epochs), and, finally, using a more recent and significantly different value (new measurements) for the decay constant of the neutron than Hayashi had used. James W. Follin Jr. of the APL/JHU was a valued collaborator in this work.

We did not immediately pursue these calculations to the point of explicitly predicting the resulting relative abundance distribution of the elements, as we had before. Such predictions were given only in some short presentations at meetings of the American Physical Society in the period 1952 to 1955. Then Herman and Alpher went on to separate positions in 1955, while Follin stayed at the APL/JHU, a situation that made it difficult to put together a detailed paper (there was no electronic mail in those days, and we were all busy with our primary tasks). Others did such calculations later, using the detailed reaction probabilities that were becoming available. We did determine that improved initial conditions still did not obviate the difficulties with the atomic weight gaps at 5 and 8. It was also clear that the initial conditions, and subsequent nucleosynthesis, would depend significantly on the neutrino types introduced into the calculation. We shall say more of this later in this chapter.

One interesting consequence of our calculations was the realization that before the onset of nucleosynthesis, all the reactions among neutrons, protons, radiation, and the other elementary particles present went on so rapidly compared with the rate of universal expansion that the mixture was basically in thermodynamic equilibrium, with abundances dependent on reaction rates rather than nuclear stability. It was also clear that the initial conditions for nucleosynthesis were dependent on one's choice of neutrino flavors: that is, the number of types of neutrinos and their antiparticles present.

We cannot resist an aside at this point. The result of much recent work on the very early universe is an inflationary model which, after a very small increment of time, is to join smoothly to the early universe description in the standard Big Bang model. If one accepts the premises of thermodynamics, however, then it must be that the state of thermodynamic equilibrium before nucleosynthesis conceals any prior history of the system. This is a well-known consequence of the application of thermodynamics, if thermodynamics is applicable to the universe as a unique system. For example, if the molecules of a gas in a closed container are all placed initially in a corner of the container, and if the molecules are then allowed to fill the container and settle down to a state with no further macroscopic changes, there is no way in principle to determine from this equilibrium state that the molecules were once confined to a special position. By analogy, if the universe went through a state of equilibrium shortly after the Big Bang (during times of the order of seconds), then it may not be sensible, except as a matter of academic interest and with little hope of observational verification, to pursue questions about the very early universe, at times of the order of 10^{-30} to 10^{-43} seconds after the Big Bang.

It may be sensible, however, to consider some phenomena associated with the earliest times in the expansion which somehow survived the equilibrium period: the present nonexistence of magnetic monopoles (i.e., freestanding "north" or "south" magnetic poles), as predicted by Dirac many years ago; the present preponderance of matter over antimatter; the existence of very small fluctuations, probably associated with the uncertainty principle, which somehow survived the equilibrium period prior to nucleosynthesis to become the anisotropies observed in the cosmic microwave background radiation by the COBE satellite and in other recent observations; and the question of how the early universe, as evidenced by COBE measurements, came to be at such a uniform temperature. (We will return to these matters later in chapter 5.)

There are other interesting consequences of the detailed examination of the state of the universe just prior to nucleosynthesis. Recall that temperature and energy of the ensemble are equivalent descriptions. Thus, an energy of one million electron volts (1 MeV) is completely equivalent to 1.16×10^{10} kelvin for the ensemble of constituents in equilibrium. When, in the expansion, temperatures were greater than the rest mass energy of baryons (protons or neutrons), baryons would be created and destroyed freely and there was purely a quark-gluon plasma.

When the temperature dropped below this value, to about 1,860 MeV, the then extant abundances of neutrons and protons would be frozen in, except for radioactive decay of the neutron. Later, when the temperature had dropped to a value of less than the equivalent of the rest mass of a pair of leptons (electron and positron)—namely, 1.022 MeV—the abundance of electrons and positrons relative to one another would have frozen in. In a similar vein, one can show that neutrinos and their antiparticles would have frozen in at a temperature whose precise value depends on whether neutrinos have a nonzero mass (a question still open in physics), as well as on the existing number of neutrino types. Because the interaction of neutrinos with other matter and radiation is extremely small, the neutrinos should then have participated in the expansion and cooling of the universe as though they were a radiative component, and there should now be a background of neutrinos which has cooled down to temperatures of the order of 2 kelvin. Their number density should be quite high, comparable to the number density of 2.725 kelvin photons, but unfortunately there seems to be no observational procedure now or waiting in the wings for detecting these very low-energy background neutrinos of primordial origin. These remarks are tentative, given that evidence of a nonzero mass for neutrinos is growing.

Two further comments should be made about our early work on nucleosynthesis. First, the very simplified model we used predicted abundances of lithium, beryllium, and boron that were high compared with observation. In 1948 we reported, with Gamow, a calculation showing that once the neutron-capture sequence was essentially finished, the temperature was still high enough so that the remaining nuclei could interact with the abundant proton nuclei present. In fact, our estimates of reaction probabilities with protons and nuclei of lithium, beryllium, and boron could have reduced the abundances of the latter by appropriate amounts. A second comment has to do with the Alpher-Follin-Herman work in 1953, as elaborated by Alpher and Herman in 1958. We raised the question about why one does not observe equal numbers of nucleons (protons and neutrons) and antinucleons in the universe and argued that the present asymmetry could not be the result of a statistical fluctuation in the early universe. At the time of our study it was accepted that the abundance of antinucleons relative to nucleons must be less than one part in 10^7, since anything greater would be able to account for all the energy generation in our galaxy or in the cosmos as a whole. Much more recent work on this question suggests that there was a basic symmetry-breaking (a kind of nonreversability) in reactions at high temperatures in which nucleons transformed into antinucleons, and vice versa, with a consequent favoring of a final abundance ratio of far less than one part in 10^7. We were taken to task by proponents of a matter-antimatter universe for not having solved this particular point, although we did suggest that the reason for the predominance of matter over antimatter arose from the physics of the early universe, which now seems to be the case.

Further Development of Nucleosynthesis Calculations

As already mentioned, we published very little on nucleosynthesis calculations after 1953; we did some further work with Follin, but the physical separation of Alpher, Follin, Gamow, and Herman beginning in 1955 precluded any detailed write-up of our results. In 1957 the seminal B^2FH paper, published by Burbidge, Burbidge, Fowler and Hoyle, explained the relative abundances of most of the heavier elements as having been generated in stellar interiors, with modification and distribution in space of the synthesized elements when the stars ran out of energy-generating nuclear fuel and went into a collapse and explosion mode. The notion that elements were made inside stars had also been put forth by a number of earlier investigators. In particular, Hoyle had suggested it in an early paper, and we had written him to raise a question that he had not considered particularly: what would happen to the abundances in the stellar interior when the star exploded and the material was processed by shock waves strong enough to induce thermonuclear reactions? We learned later through the good offices of Fowler that our written query to Hoyle on this and other matters in his work had arrived in Britain when he was receiving mailbags full of letters from listeners responding to his series of BBC talks on cosmology and was thus undoubtedly lost. The question is still under study by Stan Woosley, Craig Wheeler, Rebecca Surman and other scientists who are modeling the behavior of supernovae. Hoyle in England and Edwin Salpeter of Cornell made a significant contribution in suggesting a resonance in the probability of a reaction (later observationally confirmed by Fowler and colleagues at the California Institute of Technology) among three alpha particles as the way to get from the light elements to carbon and beyond in stellar nucleosynthesis.

The synthesis of light elements was still problematical among Steady State theorists until Hoyle and Roger Tayler, in 1964, wrote a paper entitled "The Mystery of the Helium Abundance" (*Nature 203* [Sept. 12, 1964] 1108) and used the Alpher-Follin-Herman methodology to conclude that helium must have been generated by light-element reactions in the early universe. This was a major step forward (or backward, depending on one's predilections for particular cosmological models) for these authors, since they were strong advocates of a Steady State model with no early hot, dense universe to make light elements. They also noted that the calculations were somewhat sensitive to the assumed types of neutrinos for the prenucleosynthesis era. A light-element calculation, particularly for helium, using our methodology was done by Jim (P. J. E.) Peebles in 1966, with useful results for the abundances. The first full-scale light-element calculations were carried out in 1967 by Robert Wagoner, Fowler, and Hoyle in the context of the standard Big Bang model as well as in massive stars (about 10^8 solar masses), then thought to be possible alternative sites but no longer considered a viable option.

It was interesting to us, given that we had looked at thermonuclear reactions with protons as a way of diminishing the abundance of lithium, beryllium, and boron in the postnucleosynthesis period in the early universe, when Hubert Reeves, Fowler, Hoyle, Maurice Meneguzzi, and Jean Audouze showed in 1970 that the breakup of nuclei in interstellar gas by high energy cosmic rays (almost entirely protons) would help adjust the abundances of light elements to their observed values. On the other hand, and in this context, Reeves, Audouze, Fowler, and David Schramm in 1973 could find no processes in interstellar gas which would correct the ratio of calculated to observed abundances specifically of deuterium, the major heavy isotope of hydrogen.

In 1973, Robert Wagoner carried out an improved calculation of light-element abundances, and the computer code he developed, with modern updates, is still in use by scientists in examining such abundances. As we have already noted, the one free parameter in Big Bang nucleosynthesis calculations is the extant density of matter when such reactions became important. Wagoner's work nailed down the fact that if observed deuterium abundances were primordial, then the required matter density was well below what would be needed in present observations to close the universe. Recall our earlier remarks (see the appendix also) that the standard Big Bang model, based on the Friedmann-Lemaître equations, allows for three solutions. It is now speculated that if in fact $\Omega = 1$, then there should be three contributions to Ω. One is the value of Ω associated with the requirement of primordial nucleosynthesis for baryon density, which is probably of order several tenths; we can call this Ω_b. A second contribution would be that of the yet undetermined amount of dark matter (unseen matter), which we can call Ω_m. Third, the vacuum energy density that provides the driving force for acceleration in the expansion of the universe would contribute an amount $\Omega \Lambda$ (this is related to the cosmological constant and is suggested by recent observations). We would say that the last word is not yet in on the values of these contributions to Ω.

Since the Wagoner paper of 1973, there has been a considerable effort to improve the abundance calculation for light elements. Prominent among those pursuing this work has been the late David Schramm of the University of Chicago, and Gary Steigman of Ohio State University, together with many colleagues. Reproduced in figure 3.7 is a state-of-the-art calculation of light-element abundances compared with observation in the standard Big Bang model. The vertical scale is the number density of nuclei relative to hydrogen on a logarithmic scale (except for the special scale for ^{4}He); the horizontal scale, the ratio of baryon number to photon number, is basically the combination of the density of matter and the temperature required in the universe to produce the calculated abundances (see the appendix for temperature and density relationships during primordial nucleosynthesis). What actually is plotted is the present density, since this is related by a simple formula to the density of matter and temperature at any other time (see the appendix).

Other Aspects of the Big Bang Model

We now want to address some relatively less important aspects of the Big Bang model.

Neutrinos in the early universe. We have characterized the simple Big Bang picture of nucleosynthesis as depending on one free parameter: the density of matter during nucleosynthesis. The density of radiation is uniquely defined by the model and not otherwise selected (see the appendix). As was found in the Alpher-Follin-Herman study and in subsequent developments of nucleosynthesis calculations, however, the initial conditions for this period involve the choice of the number of neutrino families. This choice somewhat affects the density history prior to and during nucleosynthesis; most modern calculations have indicated that the best representation of observed light-element abundances is obtained

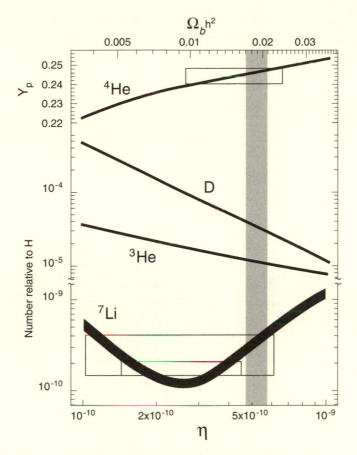

Figure 3.7. A state-of-the-art (1999) representation of the comparison of calculated and observed primordial light-element abundances. The vertical scale is in three ranges because of the large spread in values, all relative to the number of hydrogen atoms. Helium is shown on a much different scale. The several boxes delineate the range of observed abundances, including primeval helium and primordial lithium, based on two assumptions about the destruction of primordial lithium in old stars. The horizontal scale is the ratio of the number of baryons (neutrons and protons) to the number of photons, selected as initial values for the nuclear-reaction calculations. The vertical gray bar is the region of best fit to the observations, based particularly on measurements of primeval deuterium by S. Burles and D. Tytler in 1999. The scale at the top of the figure is equivalent to the bottom scale; it shows Ω_b, the ratio of baryon mass density to the density required to just close the universe, multiplied by the Hubble parameter in units of 100 km/sec/megaparsec. (Permission to use this hitherto unpublished drawing was furnished by Michael Turner and colleagues, of the University of Chicago.)

if one assumes in nature three families of neutrinos and their antiparticles. Some researchers conclude that the calculations suggest an upper limit of four families, so the matter may not be completely settled. Nevertheless, three currently appears to be a good number, and experiments at CERN (a multinational high-energy particle accelerator partly in France, partly in Switzerland, and partly in Italy) have shown that there are indeed three. With the prediction of the cosmic microwave background radiation, and the prediction that there should be three families of neutrinos, we have two cases in which cosmological prediction really preceded terrestrial observation!!

The three families of neutrinos include the electron, muon, and tau neutrino and their antiparticles—a total of six. These sundry neutrinos, stemming from different elementary reactions, differ in origin and mass —although the difference in mass is a bit of an overstatement, since experiments thus far have indicated only upper limits to their masses, as well as a more firmly founded mass difference, albeit tiny, between two of the neutrino types. There is much ferment among experimenters as this is written about whether there is a phenomenon called *neutrino oscillation*. What seems to be emerging is that one type of neutrino can convert to another type while in flight by processes requiring that at least one type has nonzero mass. It is a very exciting development, since there has for a long time been a mysterious deficit of observed neutrinos from the Sun and from supernovae explosions as compared with the calculated numbers that one would expect to observe. (The mystery may have been clarified by the time this book is published.)

If neutrinos indeed have nonzero mass, they would contribute significantly to the mean density of matter in the universe, although calculations suggest that even if they are numerous, with number densities like those of photons, they are probably not massive enough to influence the formation processes in stars and galaxies, and would not yield a high enough density of all matter to make the universe closed ($\Omega = 1$).

The initial singularity in the Big Bang. The basic equations describing the Big Bang model show a *singularity* at zero time: that is, regardless of where one is in the universe, one would find an infinite density and an infinite temperature at the origin of time ($t = 0$). Is there a simple way of avoiding this dilemma? Perhaps not, for Steven Hawking and Roger Penrose some years ago showed that a singularity was an inevitable concomitant of models of the universe, and of black holes, when such models are based on general relativity. Yet to physicists it is just as bad to say that one can divide a finite number by zero and get away with it as it is to ascribe to a physical system the existence of a singularity.

Still, there may be ways to get around it. One that we have already mentioned is to consider that early in the standard model, prior to nucleosynthesis, there is a state of equilibrium that effectively screens events at earlier times. A second is to invoke an inflationary model of the early universe in which just after $t = 0$ the universe undergoes changes from an initial vacuum state such that time, matter, and radiation come into being. We will say more of this in chapter 6. We have no doubt that the introduction of real physics in models of the very early universe will solve this problem.

The redshift controversy. It is very widely accepted that the redshift observed in the spectra of distant objects is indicative of an expansion of the universe. Although the redshifts are most easily interpreted as a Doppler shift, wherein the degree of reddening is related to the rate of expansion of the universe at the time the light emanated from the distant objects, what is actually expanding is the yardstick by which one measures distance in the universe. In other words, the expansion of the universe is an intrinsic property of the relativistic models and refers to a time-dependent dimensional change in whatever measuring tools one uses. At sufficiently large distances there is another relativistic effect: namely, the rate of clocks at the locale of emission is different from their rate at the observer, and corrections must therefore be applied to very distant redshift measurements.

The Hubble law states that redshift (velocity) is linearly related to the distance of the source. According to relativistic models, however, if one were able to look out far enough, one could detect departures from linearity, interpretable as accelerations or decelerations for closed or open models, or no departure from linearity for models just at the critical density. As we have already mentioned, the acceleration terms, or deviations from linearity, are suggested by recent observational data.

On the other hand, some astronomers challenge Hubble's interpretation, arguing that there must be some other physical phenomena producing what we observe as redshifts. The principal protagonist has been Halton C. Arp, an astronomer now in Germany who used to be at Mount Wilson–Palomar. As mentioned earlier, he claims to see instances of galaxies which, though their separation projected on the sky is very small, exhibit rather different redshifts. He concludes that the redshifts must have been produced by some process that is local to the radiating object. Arp's view is shared in a general way by a few astronomers and cosmologists who still argue that the Big Bang never happened. A detailed discussion of this controversy is beyond the scope of this book, and for most cosmologists it is a dead issue. Suffice it to say that a number of astronomers and cosmologists—for example, John Bahcall of Princeton—have looked at Arp's associations and concluded that there is a statistical problem. The number of close associations seen as projected on the sky is easily understood in terms of the scatter in the probability distribution of galactic positions across the sky.

The formation of structure in the universe.   Among the remaining problems for cosmological modeling, one that has seen much activity during our lifetimes is the origin of structure in the cosmos, where by structure we mean the formation of stars, of stellar clusters, of galaxies, of clusters of galaxies, and even possible further structure in the distribution of clusters of galaxies.

It is generally believed that fragmentation of the material in the universe began after the universe had expanded and cooled to the point that the dynamics of the expansion came under the control of matter rather than of radiation, as in the earlier universe. How much later is a matter of conjecture, although recent observations suggest that galaxy-sized clouds formed very soon after the crossover between matter and radiation control of the expansion. Gamow was particularly intrigued by this possibility. He was an early proponent of applying what is called the *Jeans criterion* (after Sir James Jeans) to the problem of forming galaxies. The physical idea is this. Suppose that at the time of equality between matter and radiation in the expansion, the universe was suffused with acoustic waves, their speeds characterized by the speed of sound at the extant temperature. (Very recent observations of the anisotropy of the cosmic microwave background are being interpreted as indicating that acoustic waves suffused the universe prior to matter-radiation decoupling.) Then there might be a wavelength among waves of successive compression and rarefaction in which the compression was enough to lead to gravitational attraction, which would in turn overcome the expansion phase of the compression-expansion wave. One can calculate the wavelength and contained mass of this incipient condensation, and it is surprisingly close to the mass and size of typical galaxies. This was an interesting idea that may again emerge as relevant.

It is not the case that there were no condensations before the radiation-matter crossover. Observations made by the COBE satellite show tiny departures from isotropy in the cosmic microwave background radiation. Since the background radiation emanates from that period in the expansion when the universe "cleared"—that is, when radiation could travel freely to distances of the order of the size of the visible universe—it appears that these departures, which are of the order of 10 parts per million, more or less randomly distributed across the sky, represent fluctuations in density or temperature in the very early universe. They were probably originally quantum in nature, originally very small, and amplified during the short period of inflation after $t = 0$, if in fact there was inflation. Then these fluctuations presumably remained as a feature of the universe to the time when the background radiation was decoupled from matter. Again, presumably, these fluctuations, whatever the explanation for their origin, provided the seeds for the development of structure, on whatever scale was dictated by the physics of processes not yet understood.

It is worth a comment that for the years of early observation when the microwave background radiation was thought to be remarkably isotropic, cosmologists were concerned about how structure could form in a medium with no "seeds" present. Several years of COBE data accumulation leading to the observation of tiny departures from isotropy removes this concern, although the transition to observed structure is not yet understood. At present, even after much theoretical and observational research, there is surprisingly little agreement about the formation of structure or even about the order of formation. As we asked earlier: did stars form first in vast gas clouds, which then condensed into galaxies and clusters of galaxies, or did galaxies or perhaps clusters of galaxies form initially as giant clouds of gas in which further gravitational condensation led to the formation of individual galaxies and stars? What is the role of black holes in galactic nuclei?

A major problem in all these matters is the now agreed-upon fact that the luminous matter by which we study objects in the heavens may be only a fraction of the matter present in these objects. (We alluded to the "missing mass" problem in our earlier discussions of Ω.) The possibility that we can see directly only a fraction of what is out there appears to have been introduced into astronomy by Fritz Zwicky of Mount Wilson–

Palomar in the 1930s. He studied clusters of galaxies, measuring the velocities of individual galaxies in the clusters. He expected these velocities to be distributed in value according to the virial theorem (i.e., characterizing a system of particles in equilibrium). To his surprise, he found that some velocities were high enough that in a finite time the individual galaxies should be able to escape the combined gravity of the cluster. He was able to calculate how long a galaxy would move about in a cluster before an interaction would boost its speed over the escape velocity, using the simple Newtonian physics theorem just mentioned, and found it to be unacceptably short, given the otherwise known ages of the clusters. He concluded that there must be a significant amount of material within the cluster which he could not see but which, by contributing to the total mass of the cluster, restrained individual galaxies from escaping.

Much more recently, Vera Rubin and Kent Ford of the Carnegie Institution studied the speed of stars in orbits around the centers of spiral galaxies as a function of the distances of the stars from the center of mass of the galaxies. (Showing that it is indeed a small world, Vera Rubin was one of a very few, including Alpher, who carried out doctoral dissertation research with George Gamow.) The velocities did not diminish with radial separation from the center as would be expected if one applied the Keplerian laws of motion to a model in which all the mass in the galaxy was luminous and could be observed directly. Rather, it appeared that there is a significant spherical halo of nonluminous mass within the confines of the stars' orbits. This was a striking demonstration of the existence (but not the identification) of missing mass, although we would mention that some spiral galaxies may have been observed not to have missing mass effects in their velocity versus radial position curves.

The nature of the missing mass is still not known. It is almost surely not all baryonic (i.e., made up of neutrons and protons), for baryons in the required densities would have participated in early nucleosynthesis in the expanding universe and would have made questionable any agreement between theory and current observation. Nor could the matter be very hot, because if it were, the chance of its being captured by gravity in galaxies and clusters of galaxies would have been quite small. Recent observations may indicate otherwise, through not yet verified by other investigators: some astronomers have found the presence of very tenuous and very hot plasma of hydrogen nuclei (and by inference helium) suffusing the universe, which could add significantly to the total of baryonic mass. The missing mass might be some form of cold matter, or it might be condensations of matter into stars that were too small to have switched on thermonuclear fusion in their interiors and would therefore now be nonluminous. Again, though, these would be baryonic, and the requirements of nucleosynthesis provide an upper limit on the number and mass of such objects. Nevertheless, astronomers are seeking such objects, and new techniques are being employed. For example, since light from a distant source passing an intervening object is deflected by the gravitational field (i.e., the photons follow a trajectory conditioned by local space curvature) of the object, one might see such deflection as a momentary brightening of the image of the distant object. The reality of this phenomenon, known as gravitational lensing or microlensing, which we mentioned earlier, has been amply demonstrated by the formation of gravitational images of distant quasars for which intervening galaxies or gas clouds act as lenses. This is all work in progress.

The cosmic microwave background radiation. One of the major features of the development of the standard Big Bang model has been the prediction and subsequent discovery of a pervasive background radiation, observed at microwave frequencies. The homogeneous, isotropic universe is filled with such radiation, blackbody in nature, at a temperature of 2.73 degrees kelvin. There is more to be said about that prediction and discovery, and we do so in chapter 5.

4

Some
Alternatives Proposed
for the Big Bang Model

Before the discovery of the expansion of the universe, many alternative models were proposed for the universe, all singularly free of real connections with observation but frequently imaginative; some were theologically based. Since the discovery of the expansion, modern alternatives have generally invoked the cosmological principle: that is, the premise that the universe is homogeneous and isotropic on a large scale. Some models originate from a philosophical predilection of the scientists for a universe that is eternal and does not change on a large scale, even if there is local evolution on a small scale. A few models try to explain the observed redshifts of distant galaxies, the principal indicators of expansion, as something other than the expansion of space; these have in general proved quite unsuccessful. And some that we mention only in passing credit the origin and evolution of the universe to an extramundane entity. We would suggest, however, that at this time in the history of cosmology the Big Bang model is, if nothing else, the paradigm against which other models must be measured.

The Steady State or Continuous Creation Theory

At about the same time (1948 and thereafter) that Gamow and the authors were pursuing the physics of the Big Bang model, Fred Hoyle, Hermann Bondi, and Thomas Gold in England were introducing a Steady State model of the universe. At least in their early work, the premises were reasonably simple. They accepted the uniformity of nature, and the seemingly irrefutable evidence for an expanding universe (although Hoyle and more recently his colleagues Geoffrey Burbidge and Jayant Narlikar have been and continue to be strong supporters of Arp's alternative to the usual redshift interpretation). They noted that the homogeneity and isotropy of the spatial distribution of matter and energy in the universe constituted a cosmological principle but went on to argue for a "perfect" cosmological principle: one espousing the idea that the universe was also homogeneous and isotropic in time. Such a principle would mean that in a large-scale view, on average, the universe was and will be just as we now observe it to be, although it would not preclude cosmological evolution on a local scale.

How then could the proponents of the Steady State explain the expansion of the universe, which seems to argue for the continued dilution of matter and energy? They could do so if some *deus ex machina* were introducing matter and energy continuously into the universe in an amount sufficient to maintain the average density at a constant value: namely, that which is now observed. The material introduced would be in the form of hydrogen and presumably helium, the principal constituents of the universe, and would be comoving—that is, moving at the common local velocity of matter—in order to blend in with the cosmic expansion. The new matter and energy would ultimately be involved in the condensation of further structures, so as to maintain unchanged the mean smeared-out density as well as the structural appearance of the universe, with new structures appearing as new galaxies between the old, if you will. In later work, Fred Hoyle and his colleague J. V. Narlikar proposed the idea of a c-field (a creation field) pervading the universe and acting as a source of the matter and energy that became part of the observable universe. It should be mentioned that the proposed rate of appearance of new matter was, and still is today, extraordinarily less than the limits of observability. This particular work presaged some of the inflationary models being considered.

As a group, Hoyle, Bondi, and Gold were vigorous and very visible in the popular press and in technical journals, and of one mind in their arguments for at least the generalities of the Steady State model. As mentioned, Hoyle and Narlikar eventually went in a somewhat different direction from Bondi and Gold, who after some years formally left the fray. Nevertheless, the general idea of a Steady State model became well known, and for a time during the 1950s and early 1960s it was widely accepted as a better paradigm than the Big Bang model. One reason for this was the major problem with the Big Bang model which we have already mentioned: that with the values for the expansion rate and mean density of the universe which were current in the late 1940s and early 1950s, the model predicted an age for the universe which was significantly less than measures for the age of the Earth and other structures in the cosmos. Another reason seems to have been philosophical on the part of those who objected to the apparently singular nature of the beginnings of the Big Bang, implying an event which is at a finite time in the past. There were many debates in the literature of the time about the oddity of substituting continuous creation of matter for a single event to explain the universe. Herbert Dingle, writing as a cosmologist in the early 1950s and drawing on the stories of Lewis Carroll, did not like either approach; he termed the argument between the Big Bang and the Steady State a choice between Tweedledee and Tweedledum.

Several scientific arguments presented problems with the Steady State model in the 1950s. One, due to radio astronomer Martin Ryle in England, was based on studies of sources of radio emissions across the sky. To attempt to put it simply, in a Steady State model Ryle expected to find the same number of sources per unit volume at all distances from the observer, correcting for the fact that there would be a bias against fainter objects at greater distance. The problem was that, on the contrary, Ryle's data ruled this possibility out and were therefore inconsistent with the Steady State model. Although Hoyle and his colleagues argued then and have continued to argue that the data were not correctly interpreted, their objections have generally not been accepted in the cosmological community.

A second problem lies simply in the cosmic abundance of helium. If the universe is in a steady state, then the presently observed relative abundances of the elements would have to be attributed to thermonuclear processes in stellar interiors, there having been no prestellar state during which light-element nucleosynthesis might have gone on. (We will not discuss some rather imaginative ideas such as "baby" Big Bangs occurring all the time and everywhere, which might lead to the observed helium abundances.) In general, processes in stellar interiors do indeed seem to account for the abundance of the heavier elements but do not appear to make enough helium, not to mention lithium, and possibly beryllium and boron. Earlier, we mentioned a major publication in 1957 on the synthesis of elements in stars, the so-called B²FH paper. Early in 1999, Hoyle, Geoffrey Burbidge, and Narlikar published an argument that sufficient helium *could* be made in stellar interiors, but they failed to detail a mechanism. (This argument and others by this team were originally published in *Nature* and in *Physics Today* in 1999; in 2000, they have authored a book, which we have not yet seen.)

A key question in stellar nucleosynthesis was how the light elements built up toward the heavier elements; there was an apparent barrier to the formation of carbon and subsequent heavier species. A theoretical solution to this problem was suggested, as already mentioned independently by Hoyle and by Edwin Salpeter at Cornell University. There was an observational verification by Fowler and his colleagues at CalTech of the probability for the formation of a carbon nucleus by a three-body reaction, the three bodies being helium nuclei (i.e., alpha particles), leading to an excited state of the ^{12}C nucleus. Such a resonance or enhanced-reaction probability makes it possible to continue the reaction sequences leading to the heavier elements.

A third problem with the Steady State model lies in the observation of the relict 2.7 kelvin cosmic microwave background radiation, first predicted by Alpher and Herman in 1948 and definitively observed by Arno Penzias and Robert Wilson, of Bell Telephone Laboratories, in 1965 (see chapter 5). Hoyle in particular, with Burbidge and Narlikar, has tried to rationalize this radiation as arising from the thermalization of starlight by interstellar and intergalactic dust grains. But that explanation requires the existence of a rather special distribution of grains (suggested as being principally iron), which appears to be quite unnecessary, given the simplicity of the explanation for this radiation in the Big Bang model. Moreover, there is no evidence for the existence of such grains, particularly as their interaction with intergalactic radiation might affect the appearance of distant objects.

The recommended reading list at the end of this book includes an eminent historical account of the Steady State versus Big Bang controversy published in 1996 by Helge Kragh, a Danish historian of science. With a few holdouts, the Steady State model has become solely a historical entity, although it has served to sharpen up thinking on the Big Bang paradigm. Moreover, Hoyle, Burbidge, and Narlikar continue to do battle.

Redshift as a Local Phenomenon

Hoyle and a few other astronomers and cosmologists continue to be skeptical of the validity of the Big Bang model. One of these is Arp, now at the Max Planck Institute for Astronomy in Germany and formerly at the Carnegie Observatories, who has some very limited support for his suggestion that redshifts are developed locally in the sources of radiation. Among his supporters in at least a general way are Hoyle and, again, Geoffrey Burbidge. As noted in the discussion of Arp's skepticism in chapter 3, there is no physical evidence for such a local effect.

Matter-Antimatter Universe

Another scientist who had a rather dour view of the Big Bang was Hannes Alfvén, well known for his pioneering research on the physics of interaction of electrically conducting gases and fluids with any magnetic field present. Alfvén and another scientist, Oskar Klein, argued that the universe must reflect the symmetry in nature between the existence of matter and antimatter. Alfvén wrote a short book on the consequences of this symmetry, with the imaginative title *Worlds-Antiworlds* (1966). He and Klein recognized that matter and antimatter could not survive for long if they were intimately mixed, since the particles and antiparticles would mutually annihilate and release large amounts of energy in the form of radiation. It is well known, observationally and theoretically, that if a particle and its antiparticle meet, annihilation will ensue, with the sum of kinetic energies of the particles and antiparticles, as well as the energy equivalent of their masses being carried off as electromagnetic radiation by two photons. We mentioned earlier that as little as one part in 10^7 of antimatter intermixed in galaxies could easily account for the entire energy output of such objects if they contained equal amounts of matter and antimatter.

What Alfvén and Klein proposed was to abandon the cosmological principle: to suppose that the universe was not homogeneous and that there were in the cosmos objects which on a large scale—say, a galactic scale—were made of either matter or antimatter. It is not at all obvious that we could distinguish observationally between a galaxy made entirely of antimatter, and one made entirely of matter. To the best of our knowledge, the radiation from an antimatter galaxy, as well as its gravitational effects on other objects nearby, would appear just the same as what we now observe and attribute to objects made of normal matter. Should a matter and an antimatter galaxy meet, of course, there would be violent destruction and the creation of intense radiation. To ameliorate this result, they hypothesized that at the interface during interactions there would be a region of *leidenfrost*, much like the layer of vapor between a hot pan and a drop of water, which dances around on the hot surface. Such regions of intense radiation should be observable, but none has been identified.

The notion of a matter-antimatter model of the universe has almost dropped out of consideration, given the finding that in the reactions which would normally maintain an ensemble of particles and antiparticles in equilibrium in the early universe, there is an asymmetry between the nuclear reactions "particles going to antiparticles" and the reverse reactions "antiparticles going to particles." Theoretical calculations in what have come to be called *grand unified theories* suggest that the asymmetry is consistent with the observed great preponderance of one type with respect to the other. Such theories have been studied as a way of understanding that in the early universe, or in experiments at energies not yet attained in the laboratory, the several forces of nature—nuclear, electromagnetic, weak, and gravitational—all had the same magnitude. But grand unified theories have not yet succeeded in integrating gravitation into the unification. Thus far electromagnetic and weak have been unified.

As recently as October 1997, however, some new papers appeared in the peer-reviewed literature proposed several possibilities for the origin of the small departures from isotropy in the cosmic microwave background radiation. One in particular suggests that they may be the result of interactions between matter and antimatter regions in the early universe, with the departures in temperature reflecting matter-antimatter annihilation at the interfaces of the several regions. Alternatively, it has been proposed that the properties of these small departures reflect the existence of *cosmic strings* in the early universe. One view of cosmic strings is that they are topological defects (tears in the "fabric" of spacetime) in the geometry of the early universe, defects that survived inflation to become the large-scale structures we now see.

We had some correspondence with Hannes Alfvén in the mid-1970s; he was concerned that we and other scientists working with the Big Bang model did not seem to be taking his collaborative work with Klein on cosmological matter-antimatter asymmetry all that seriously. In particular, he faulted us for not having properly discussed the matter-antimatter problem or dilemma within the context of the Big Bang model. We had in fact not ignored this problem but did not know how to deal with it in any detail at the time these criticisms were made. We did calculate how much antimatter should be left vis-à-vis matter on the basis of some statistical arguments and concluded that it would be a very small number (like the mass of the earth compared with the mass in the visible universe). And in at least one paper we opined, as mentioned earlier, that the matter-antimatter asymmetry question would perhaps be solved by the physics of the very early universe prior to nucleosynthesis, as indeed it appears to have been, at least qualitatively.

Alfvén's other arguments against the Big Bang have been dealt with over time. For example, he questioned the blackbody nature of the background radiation, first because the initial observation of Penzias and Wilson was at a single wavelength, not proving that the radiation was blackbody. He was likewise critical when some data were reported (later shown to be an experimental artifact) at other wavelengths which were again inconsistent with the blackbody spectral identification. The measurements of the background radiation provided by the COBE satellite have put this criticism to rest. He thought that primeval nucleosynthesis was not necessary because helium could be made inside stars. As it turns out, that still does not appear to be possible. Yet another objection was his concern for stellar modeling, because terrestrial observations of neutrino flux from the Sun do not match the calculations for this flux based on thermonuclear reactions in the solar core. This is indeed still a problem, although recent work on neutrino oscillations and neutrino mass, mentioned in chapter 3, may finally resolve it.

As a consequence of Alfvén's studies of cosmic electrodynamics (the interaction of moving, electrically conducting fluids in the presence of pervasive magnetic fields), there is a considerable body of work attempting to model the principal observed features of the cosmos using the fact that most of the matter is in a plasma state, with behavior controlled by forces acting on plasmas in the presence of magnetic fields rather than by gravitational forces. Although the existence of a plasma state and associated electromagnetic forces is not arguable, some of the rationale for going to such a description has been negated over time. For example, that the cosmic microwave background radiation is indeed blackbody in nature has been well verified by ground-based measurements since the discovery of the radiation by Penzias and Wilson, and particularly by the remarkable data set from the COBE satellite. Other facets of the plasma model involve an inhomogeneous universe, for which there is no evidence on a large scale, as demonstrated, again, by the COBE results and other studies.

A final comment on the blackbody nature of the background radiation. It has been suggested that in the plasma which is the pervasive state of the early universe thermal and collisional effects can cause deviations from a pure blackbody spectrum at low frequencies, frequencies below those characteristic of plasma (which have vibrational modes of excitation suggesting a gelatinlike material). The calculation suggests a net effect during the period of light-element nucleosynthesis of the order of about a fraction of a percent in the density of radiation at the time, and, if confirmed, would probably alter the detailed calculations of light-element abundances by an extremely small amount. As of this writing, all such conjecture is very much in the nature of fine-tuning and work in progress.

The "Constants of Nature"

One body of scientific speculation about cosmology probably had its modern origin in some publications by the physicist P.A.M. Dirac, known for his basic work in quantum mechanics. In 1937—when, it is reported, he was on his honeymoon—Dirac wrote several papers in which he postulated a "law of large numbers." He pointed out that one could form appropriate dimensionless ratios of various quantities in atomic and nuclear physics, and in cosmology, whose numerical values were of the order of $10^{(40)n}$, where n takes on values of the order 0, 1/2, 1, 3/2, 2, and so on. He also noted that 10^{40} was closely related to the number of particles in the visible universe, $10^{2(40)} = 10^{80}$, the so-called Eddington number, which Sir Arthur Eddington introduced into cosmology. Obtain the number of particles in the Sun, say, by dividing the mass of the Sun by the mass of a proton or neutron, and then multiply the result by the number of visible stars. That number is obtained by assuming a typical number of stars in a galaxy—say, 10^9—and multiplying by the estimated number of visible galaxies inside a sphere of radius ct_{age} (the velocity of light times the age of the universe). In one of Dirac's dimensionless ratios involving time, he used as a characteristic time the age of the universe. If the universe is evolving, as he thought it was and which certainly appears to be the case, then maintaining his several ratios as time-invariant required that the constant of gravitation vary with the epoch.

In retrospect, it is amazing that Dirac's so-called law of large numbers was so quickly adopted by a number of cosmologists in their attempts to modify extant models of the universe. In particular, since Dirac concluded that the gravitational constant G was not a constant but, rather, a function of the epoch, a number of models were constructed which involved a time-dependent G. Perhaps the most highly developed models were those of Pascal Jordan and colleagues in Germany and of Carl Brans and Robert Dicke at Princeton. In the Brans-Dicke theory, really put forth as an alternative to the general theory of relativity, the concept of a time-dependent gravitational constant is not dependent on Dirac's law of large numbers. Observational cosmology has put a limit to the mathematical expression of this time dependence, namely, it restricts the applicability of the Brans-Dicke theory to models that are essentially flat—that is, exhibiting Newtonian behavior. The Brans-Dicke model may have relevance to some inflationary models of the very early universe, but the work of Jordan and others on a time-dependent gravitational constant in cosmology has not stood up to the test of time and observation. We should mention that in one of the last pieces of work Alpher did with Gamow before Gamow died, we suggested that in the spirit of dimensional analysis, one should not take the age of the universe as a characteristic time for the universe; instead, one should use an invariant characteristic time. We noted that in the expansion in the standard Big Bang model, the time when matter and radiation densities were equal (radiation controlling the expansion before, and matter controlling the expansion after) was actually a constant. Using this enabled us to write all the dimensionless ratios proposed by Dirac as 10^{36n} instead of 10^{40n} and led to a model with a constant value of the gravitational constant G. Dirac was not happy about this result but suggested that observation in the end would settle the question—and observations, including time delay measurements for signals from the Mariner mission to Mars, have indeed ruled out a time-dependent G.

Another theory that attempts to replace general relativity (due to Hoyle and Narlikar) has the interesting feature that in this model particle masses are time-dependent. A consequence is that the variation of masses is in turn responsible for redshifts in galactic spectra. In this approach the motivation is the fact that there have been no observational checks on the theory of relativity in the presence of strong gravitational forces. Otherwise, the model tracks general relativity satisfactorily but has not yet proved to be a useful approach to cosmological modeling.

Theories based on a time-dependent constant of gravitation G or its equivalent have gotten good play from other cosmologists, as they were serious attempts to construct useful models. But observations overtook them and they have essentially faded from further consideration.

The Anthropic Principle

This is perhaps the logical place to introduce the anthropic cosmological principle (but, see also chapter 9). For those who are interested, there is a 1986 book by that title, dealing with cosmology generally and with this somewhat exotic corner of it, by the English astronomer John Barrow and Frank Tipler of Louisiana State University (see our recommended reading list). We mention it here because of the use made of the constants of nature in developing the anthropic principle. Proponents of this view usually take the Dirac law of large numbers very seriously.

To put it briefly, the argument is that there must be a connection between the physical properties of the universe and the existence of intelligent life within it. If those properties were different, we would not be here trying to explain the structure and evolution of the universe. For example, the density ratio Ω—that is, the ratio of mean cosmic density to the critical density—is so close to unity that one might regard it as somehow having been set up for the existence of life. Our view, simply put, is that evolutionary pressures have brought us to where we are, in compliance with the constraints put upon us and other forms of life by the universe in which we live. We suggest that life has evolved to fit the constraints, rather than the other way around. Nevertheless, the arguments for an anthropic principle are provocative, and we urge the interested reader to have a look at Barrow and Tipler's book as well as our somewhat expanded discussion in chapter 9.

Let us consider the possibility that others among the constants of nature may be functions of the epoch: for example, the unit of charge on elementary particles, the velocity of light, the constant of gravitation, Planck's constant (which expresses the proportionality between the energy and frequency of photons), various elementary particle masses, and the fine-structure constant (a numerical constant that is important in atomic physics). As we have said, a proposed *weak* anthropic principle assigns humankind a special place in the universe. Human beings have a number of elements in their composition which according to present views must have been synthesized in stellar interiors, then distributed in space for further evolution of cosmic bodies and, not so incidentally, ultimately used in the makeup of life forms; it is by now a cliché that all the material in life forms on Earth is "starstuff," a description which may have originated with the late Carl Sagan. The processes that form elements in stellar cores depend on the constants of nature, and these have values consistent with a lifetime of stars sufficient to have allowed humankind to evolve. The weak principle merely requires lifetimes of the order of billions of years to be consistent with the development of life.

Then there is a *strong* anthropic principle, which notes that the constants of nature appear to be finely tuned to values that permit the development of intelligent life. Those who take this principle seriously argue that there must have been a creator who set up a cosmos in which these constants were fine-tuned to allow life. Some argue that there are many universes, and that we have developed in the one that permits life as we know it. The idea of many parallel universes is interesting, but there is no evidence for its validity.

Again, the counterargument we suggest is simply that life evolved in our universe in concert with the constants of nature as we find them. Had there been other values for these numerical constants, either life would not have evolved in the universe in which we are situated or a form of life different from ours and consistent with these constraints would have evolved.

Other Models

We have over the years accumulated a number of books and several files full of articles and correspondence whose authors—sometimes scientifically trained, albeit not always in the disciplines relevant to cosmological modeling—offer well-meaning alternatives to the Big Bang. This book is not meant to be in any sense a history of these alternatives. We apologize to authors not mentioned for not dealing in any depth with the idea of a plasma universe, or universes in which particles are little vortices, or a universe with two time scales that would obviate the philosophical problem of an origin of the cosmos at a finite time in the past. At least one author raised the question of difference between time measured in atomic systems (with the time interval related to nuclear or electromagnetic forces and having nothing to do with gravitational forces) and time measured in terms of the motion of heavenly bodies (based on gravitational forces). As it turns out, this difference has been proved nonexistent by a number of observations, including, for example, the delay time for signals from the Mariner probe to Mars as the radio signals generated by an atomic system aboard the probe traverse the gravitational field of the sun.

Some of the proposed alternatives for the Big Bang exhibit a lack of recognition of existing observational cosmology and its consequences. Others are more scholarly efforts to push the envelope, as it were, of theoretical approaches to cosmological phenomena. We have limited our attention to a few models whose authors do recognize the need for observational verification if a theory is to be considered; these all help to sharpen up ideas within Big Bang cosmology. One attribute of the scientific process is the ever present possibility that a new theoretical approach or a new set of observations will send a given model of nature into eclipse and require a new approach, or at least lead to major changes in an existing model. The approaches described here, however, simply do not do as well as the Big Bang model.

5

The
Cosmic Microwave
Background Radiation

Everyone agrees that 1965 was an important year in the historical development of cosmology; indeed, some take it as the birth year of modern cosmology. The book *Origins*, by Alan Lightman and Roberta Brawer (see our recommended reading) consists of interviews with the authors' choices of leading cosmologists from their perspective of 1990. In reviewing the book, Geoffrey Burbidge, a noted astronomer and cosmologist, and a leading critic of the Big Bang model, listed among other serious deficiencies the fact that the authors seem to consider 1965 the starting year for cosmology. Lightman and Brawer are not the only authors who think that. Why is this so?

Certainly it was a pivotal year. In 1964, satellite communications were entering a period of rapid development, and researchers were interested in improving their ability to provide clean signals for commercial exploitation of this new technology. Thus it was that for some years before 1965, Arno Penzias and Robert Wilson of the Bell Telephone Laboratories had been conducting studies to develop a microwave receiver that would have as low a noise figure as they could manage, with the ultimate goal of having the best possible receiver for satellite communications. Their test body at the time was the ECHO satellite, basically a large, metallized Mylar balloon in orbit. After very carefully eliminating all the sources of interference they could imagine (some were almost unimaginable: e.g., pigeon droppings in the horn of their receiver), these investigators found that they were left with a strange signal of unknown origin. It was in fact a background radiation in the microwave region of the electromagnetic spectrum, at a wavelength of 7.3 centimeters, which appeared to be unpolarized and isotropic—that is, independent of the direction of observation—and with an intensity in their first measurements corresponding to about 3.5 kelvin if the source was assumed to be a blackbody. We digress here for a brief and somewhat technical description of blackbody radiation, since it plays so important a role now in cosmology.

A blackbody is a physical system that emits thermal radiation and whose spectrum (i.e., the distribution of radiated intensity as a function of wavelength) is described as displaying a Planck distribution, named after the German physicist Max Planck, who first proposed it early in the twentieth century. The main feature of such radiation is that the form of the spectrum, as well as its intensity scale, is completely specified by a single number: namely, the value of a characteristic temperature. The radiation output of the Sun and most stars is blackbody in nature, although the overall intensity of their Planck-like spectrum is altered by the dropoff of intensity according to the inverse square of the distance from the radiating source to the terrestrial observer. If one measures the absolute intensity at one wavelength, and the source is blackbody, then the spectral distribution and the characteristic temperature are completely specified. To their credit, Penzias and Wilson—who report that at first they did not know they were making a fundamental cosmological observation—were sufficiently confident of their observing techniques not to dismiss the observation as spurious but rather to accept it as something to be understood.

The Planck nature of the spectrum has been thoroughly confirmed over the years since then. The most beautiful confirmation was made recently by the COBE satellite (of which more later in this chapter). The most immediate confirmation was made at Princeton shortly after the Penzias-Wilson observation by Peter G. Roll and David T. Wilkinson, using a Dicke radiometer (an instrument developed by R. H. Dicke of Princeton), which had been nearly ready to go when the Bell Laboratories story broke. The Princeton group was in effect scooped. Figure 5.1 shows the theoretical Planck spectrum for several temperatures. Over the years there have been many observations verifying the properties of the cosmic microwave background radiations (CMBR). Many of these have been direct observations, using the same approach as Penzias and Wilson. There were in fact a number of observations made prior to those of Penzias and Wilson, whose significance was not recognized, as pointed out below.

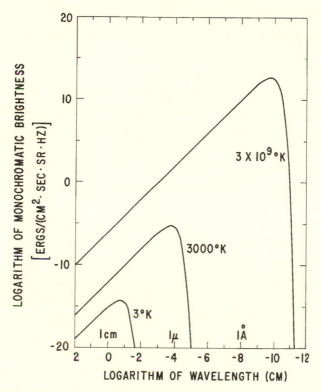

Figure 5.1. The theoretical spectrum of Planck blackbody radiation for three temperatures. The vertical scale is relative brightness at a specific wavelength, and the horizontal scale is wavelength in centimeters. Note that 1 μ (micron) is 10^{-4} cm, and 1 Å (Ångstrom) is 10^{-8} cm.

The Impact of the 1965 Discovery

This radiation observation reported by Penzias and Wilson in 1965, and its immediate interpretation (in an accompanying paper by Dicke, Peebles, Roll, and Wilkinson, all at Princeton) as the relic radiation from an early hot, dense phase of an expanding universe, has had a profound effect on our world picture. Steven Weinberg, in his book *The First Three Minutes* (1977, revised 1988) has an exciting dramatization of the observation and its interpretation. It now seems clear that for some time before 1965 the late physicist Robert Dicke at Princeton (as reported by his long-time colleague Peebles in a 1971 book, *Physical Cosmology* [Princeton: Princeton University Press], and elsewhere) had wondered if the universe might in fact be modeled as oscillating, as a universe that had gone through a recurring series of expansions and contractions. Such a model is within the allowable framework of relativistic cosmological models and had been well studied mathematically.

We must digress here to remark on the question of entropy in the universe. The entropy question has been considered by a number of investigators, recently including Alpher and George Marx. During the expansion, matter (limiting our consideration to the smeared-out density of matter) and radiation follow a different dependence on temperature. Heat transfer between matter and radiation depends not only on the square of the difference in temperature but also particularly on the nature of matter-radiation interactions. In the early universe, such interactions would have been fast compared with the rate of expansion, and the two components would have been in thermodynamic equilibrium. Today, the temperatures of matter and radiation (ignoring structure formation) would be fractions of a degree of the order of 0.003 kelvin and 2.725 kelvin, respectively), with a corresponding growth of entropy during the expansion. The entropy production, expressed in terms of radiation entropy per baryon in a very simplified model, is of the order of 10^{10}, which agrees with the value needed for nucleosynthesis.

If the universe is closed, and going through a series of expansions and contractions as time goes on, then one finds that the model emerges from each compression and proceeds to a higher entropy state. The primary determinants for the model's being termed closed and therefore capable of exhibiting oscillatory behavior are the mean density of matter in the universe and the expansion rate. At the time of this writing these measures are still not well enough determined to distinguish between models that are closed and ultimately stop expanding and return upon themselves, models that will continue to expand indefinitely and are called open, and models whose mean density is exactly equal to a critical density that divides the set of possible open universes from the set of possible closed universes. As discussed in chapter 6, since 1965 there has been much work on an inflationary beginning for the universe, which, its proponents argue, requires a cosmic mean density precisely equal to the critical density, for all time.

Dicke speculated that if the universe is oscillatory, then it would follow that at a time associated with maximum compression in such a model, a very high temperature and density would have served to wipe out all previous history and to allow a new cosmic composition to be established. Given the extreme physical conditions associated with such compression, he proposed the possibility that residual radiation from this early hot, dense phase might now be detectable. By early 1965, Dicke and his colleagues Roll and Wilkinson were engaged in building a microwave radiometer of Dicke's design to look for this radiation. In 1946, Dicke and others had built such an instrument to study radio emission from the atmosphere of Earth and had concluded that there was a residual radiation component which they could neither eliminate nor explain and which corresponded to a temperature of about 20 kelvin if the unknown source was blackbody. No cosmological connection was made between that measurement and the later search for a residual radiation from the early stages of an oscillating universe until after Dicke and his colleagues learned of the observations of Penzias and Wilson.

Meanwhile, in 1964 Dicke had asked his Princeton colleague, Peebles, then in a postdoctoral position, to calculate the observable characteristics of the radiation. In 1965, Peebles gave a talk on his calculations (he suggested a present background temperature of about 10 kelvin) at the Applied Physics Laboratory of Johns Hopkins University, where we had done our early calculations of background radiation. But no one among our former colleagues at Johns Hopkins who might have recalled our work was present at the lecture. We had long since departed, and Follin, who had worked with us around 1953, was unaware of the talk. Meanwhile, Bernard Burke, a radio astronomer at MIT who had heard of the talk and may even have had a preprint of Peebles's calculation, got word to Penzias and Wilson, who contacted the Princeton group. When the Princeton researchers thus learned of the Bell Laboratories result prior to its publication, they identified the 3.5 kelvin result as the relic radiation from an early hot, dense phase of an expanding and evolving universe. This identification by Dicke, Peebles, Roll, and Wilkinson was published in July 1965 in the *Astrophysical Journal*, back to back with the paper announcing the observation of background radiation by Penzias and Wilson. For a time in subsequent publications the Princeton group described this background radiation as arising from a "primeval fireball"; later, the name Big Bang, coined earlier by Hoyle, stuck. (Of historic interest, perhaps, is the fact that as early as 1928, Eddington had objected to the notion that the universe began with a bang, although he did not refer to it as big. And especially vivid are the words of Georges Lemaître on page 78 of his book 1950 *The Primeval Atom: An Essay on Cosmogony*: "The evolution of the world can be compared to a display of fireworks that has just ended: some few red wisps, ashes and smoke. Standing on a well-chilled cinder, we see the slow fading of the suns, and we try to recall the vanished brilliance of the origin of the worlds." Not bad.)

Shortly after the Penzias-Wilson observation, Roll and Wilkinson at Princeton got their radiometer running and confirmed a 3 kelvin background at another wavelength: namely, 3.2 centimeters. It was the first among a large number of confirmations of the existence and temperature of this radiation. The NASA satellite called COBE provided stunning data establishing that the microwave background radiation shows a beautiful Planck spectrum at a temperature of 2.725 ± 0.04 kelvin (i.e., an uncertainty at the 95% confidence level) and that this radiation is unpolarized and isotropic to better than one part in 100,000, as shown in a very recent representation of the results shortly before the COBE instrument was turned off after four years (figure 5.2). When John Mather of NASA's Goddard Space Flight Center, project scientist for the overall COBE project and principal investigator for one of the three instrument teams (namely, the background radiation spectral measurement), presented an earlier version of the spectrum determined by COBE after a few weeks of data collection, during a talk to a crowded lecture hall at the meeting of the American Astronomical Society held in Washington, D.C., in January 1990, he and the instrument and the result received a standing ovation. Details on the COBE satellite and its instrumentation, as well as insights into the human efforts that went into the implementation of the four-year program, may be found in a 1996 book by John Mather and a science writer colleague, John Boslough (see our recommended reading list). On a personal note, we were delighted to have been invited to speak about our 1948 prediction of a cosmic background radiation to a group of scientists and science writers assembled to observe the launch of COBE, and we were privileged to have a VIP tour of the COBE vehicle on its launching pad as well as to observe its launch.

What is the origin of this radiation? The concept of the early universe in the Big Bang model is that after a very brief time interval, during which an inflationary start perhaps explains the cosmic behavior, the universe was in a highly homogeneous and isotropic state, expanding and cooling, with the only structure being some departures from isotropy on a scale of 10 parts per million, exhibited when the background radiation emerged from its epoch of last scattering. (We will say more of these departures later. Some inflationary theorists argue that the early universe may in fact have been inhomogeneous, which would account for the departures from isotropy.) For quite a long time after the Big Bang—

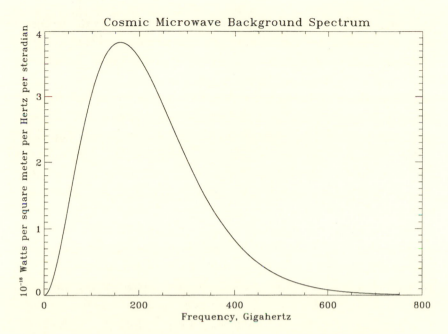

Figure 5.2. The spectrum of the cosmic microwave background radiation, shown as observed signal strength versus observing frequency, as determined by four years of measurements by the Cosmic Microwave Background Explorer (COBE) satellite and based on literally millions of data points. The error bars on the measurements are not shown because they are actually smaller than the width of the plotted curve, which gives the theoretical spectrum for a 2.725 kelvin blackbody. (This drawing was supplied by NASA, the Goddard Space Flight Center, the COBE Science Team, and in particular John Mather, the principal investigator for COBE.)

say, half a billion years—the material of the universe was in a plasma state, containing hydrogen and helium nuclei and electrons, and was quite opaque to the passage of the ambient thermal radiation. (Parenthetically, we remark the suggestion of some recent studies that the interaction of matter and the background radiation may have influenced the density of the cosmos at the time of light-element nucleosynthesis at no more than the 1% level). The radiation was in effect trapped, being absorbed and reemitted over distances comparable to the distances between the free nuclei (hydrogen and helium nuclei principally) and electrons. When cooling had dropped the temperature to several thousand degrees, recombination began to occur (i.e., electrons became attached to the nuclei present), so that the gas became a gas of neutral atoms and in a short time changed from being opaque to being transparent to the ambient thermal radiation. After this time the radiation could travel quite freely, on distance scales comparable to the size of the visible universe (had there been someone available at the time to look around).

It is this freely traveling radiation, which began at a temperature of several thousand degrees, that has by now cooled in the cosmic expansion to the presently observed temperature of 2.725 kelvin. The high level of homogeneity and isotropy exhibited by the radiation is a reflection of the homogeneity and isotropy of the distribution of matter at the time of recombination or decoupling. As already mentioned, departures from isotropy, observed to be of the order of 10 parts per million, therefore tell us something about the density anisotropy of the matter distribution at decoupling. What COBE actually observed as anisotropies, on an instrument whose operation and results have been described in a 1993 book by team leader George Smoot and science writer Keay Davidson (see our recommended reading), were excursions in temperature from the basic 2.725 kelvin background, which can be interpreted as excursions in matter density, perhaps due to acoustic waves. We will say more about the anisotropy data later in this chapter.

Thus far, to an outside observer the story seems more or less unexceptional. An excellent and original observation was reported by Penzias and Wilson, and another group, Dicke, Peebles, Roll, and Wilkinson, was in a position to provide a rational explanation for it. By late 1965, however, the importance of the observation of a cosmic background radiation was becoming widely recognized as a major demonstration of the validity of the Big Bang model, as was the success of the model in obtaining theoretical agreement with observations of the cosmic helium abundance. In fact, given the flood of papers that emerged during subsequent years, we and Gamow became concerned that our work in developing the Big Bang model, in predicting the background radiation, and in rationalizing the high cosmic helium abundance had become lost. At the suggestion of Gamow, therefore, we joined him in writing a brief review of all this for publication in 1967 in the *Proceedings of the National Academy of Science*, a journal noted for fast publication of papers by members but unfortunately not widely read.

We have been asked why we did not pursue the possibility of the observation of the cosmic microwave background between 1948 and 1965, when the Penzias-Wilson results burst upon the scene. Had anyone recognized the experimental possibilities? During this period, as already mentioned, we had explored the possibility with radio astronomers and others to no avail. Perhaps our earlier prediction was not taken seriously, as indeed was suggested by Weinberg in *The First Three Minutes*.

Prior to 1965, however, there were at least eight literature references in peer-reviewed archival journals which referred in one way or another to the prediction of a cosmic background radiation. These included some papers by Gamow after he accepted, in 1950, the reality of the 1948 prediction by Alpher and Herman. There was in addition a paper in a Soviet journal by Andrei Doroshkevich and Igor Novikov, submitted by Yakov Z'eldovich and published in 1964, which made reference to a 1949 paper by Gamow: "According to the Gamow theory, at the present time it should be possible to observe equilibrium Planck radiation with a temperature of 1–10 kelvin" (p. 112). We cannot fathom this statement, for the 1949 Gamow paper makes no mention whatsoever either of our prediction or of anything that might be interpreted as referring to the present value of the predicted background temperature. There is similar confusion in a separate paper by the Russian physicist Z'eldovich (in *Soviet Physics—Doklody* 9[2] [August 1964]), which discussed the Gamow proposition that there might be a background radiation but concluded, on the basis of an erroneous interpretation of work published by Edward Ohm in the *Bell Technical Journal* (studies using the same antenna as Penzias and Wilson), that any background would have a temperature less than 1 kelvin; that would rule out the Gamow result, which, however, Z'eldovich mistakenly thought was much higher (25 kelvin). These matters are discussed at length by Kragh in *Cosmology and Controversy* (p. 345).

Early Measurements of the CMBR

The tangled history of the discovery and interpretation of the cosmic microwave background includes some measurements made before 1965. One incident involving Nicholas Woolf concerned the early work of Walter S. Adams and Andrew McKellar on the rotational temperature of the CN radical (cyanogen) as seen in absorption in interstellar space.

The CN radical is a simple diatomic molecule, made up of a carbon atom and a nitrogen atom, configured something like a dumbbell; as such, it can rotate about an axis perpendicular to an imaginary line joining the two atoms. Such rotation is a way for the molecule to store energy, and the storage is subject to being quantized: that is, the molecule has specific states in which the energy can be stored as rotational energy. Thus, spectroscopists who study these matters find that there is a ground state of lowest energy, a first excited state storing a discrete amount of energy, a second excited state storing yet more energy in a discrete state, and so on. If the molecule is in equilibrium with a radiation field—that is, a gas of photons with a Planck energy spectrum—that constitutes a thermal bath, then the number of molecules in the ground state, in the first excited state, and so on, will reflect the temperature of this bath. The population of the various states depend on their energy above the ground state and their multiplicity, a measure of the complexity of the particular rotational configuration (i.e., the number of different ways in which the particular state can store energy). If now we look toward a luminous object, the intervening CN molecules will absorb energy from the light of the object at each of the rotational energy levels in which the molecules find themselves, the amount depending on the population of the molecules in each of the states. Thus, if we see absorption at wavelengths corresponding to the several rotational states, we can infer from the amount of absorption in each state the temperature of the photon gas that constitutes the radiation bath in which the molecules are resident. To calculate populations from observed intensities of absorption in various states requires that one know *oscillator strengths* for these states, which are a measure of their multiplicity. Oscillator strengths were not known very well for CN before 1965.

To continue the connection with Woolf: he had discussions with George Field (now of the Smithsonian Astrophysical Observatory; both Field and Woolf were at Princeton in 1965) about the possibility of finding a radiation field in space by studying the absorption of radiation by interstellar molecules. Field suggested that some other phenomenon might explain such observations. He prepared a manuscript on the subject but because he did not know the properties of the several rotational states in CN, he put his paper in a drawer.

After Woolf had left Princeton and was at the University of Texas–Austin, Dicke got in touch with him through William Hoffman (also then at Texas) to ask Woolf to try a microwave measurement of the background radiation during a balloon flight. Thinking it would be more reasonable to look for temperature measurements involving interstellar molecules, Woolf reported to Hoffman and Patrick Thaddeus, then at NASA's Goddard Institute for Space Studies in New York, that W. S. Adams had given a prescription in 1940 for obtaining temperatures if one could observe and interpret transitions among rotational levels. Thaddeus looked at Adams's paper and found a brief mention that McKellar in the same year had indeed observed absorption intensities for CN rotational lines. On the basis of his data, McKellar suggested an upper limit to the effective temperature of interstellar space ranging from 0.8 to 2.7 kelvin, the uncertainties being the consequence of not knowing the relative probabilities of the several rotational states. In fact, he proposed that the result was 2.3 kelvin, a remarkable result indeed!

Woolf also reports having heard that Iosif S. Shlovskii at about the same time had made a similar suggestion about rotational temperatures in a colloquium given at the Sternberg Astrophysical Institute in the USSR. Neither Woolf nor we have found a literature reference to this talk, so it remains hearsay. Nevertheless, it appears that had everyone been studying the literature in a careful way, the existence of the background radiation would have been realized well before the Penzias-Wilson observation. When all is said and done, the 1965 observation of the CMBR and its interpretation were and are unequivocal.

This does not exhaust the list of suggestions in the published literature, understandable in retrospect, of measurements or calculations that could have been interpreted at the time as indicating a CMBR. Perhaps the earliest was that of Sir Arthur Eddington, who in 1928 (*The Nature of the Physical World*) pointed out that the radiant energy from all stars that were resident in space implied a total energy density corresponding to a temperature of 3.2 kelvin. Eddington said that a blackbody in space should assume such a temperature but did not indicate a way to make a measurement. There was of course no cosmological implication contained in Eddington's suggestion.

Several actual measurements were made in the mid-1950s. A Russian radio astronomer, Tigran A. Shmaonov, used a captured German radar antenna to study the background radiation from the sky and found a result of several degrees kelvin. He reported his observational technique in an obscure Russian journal on instrumentation but did not make any cosmological connection until many years later (after Penzias and Wilson) in a colloquium reported by Igor Novikov. Also in the 1950s, a French radio astronomer, Emile Le Roux, obtained a result of about 3 kelvin from measurements at 33 centimeters, again with no cosmological inference. Gerhard Herzberg (Nobel in Chemistry), in his classical text on molecular spectra and molecular structure, published in 1950, discussed the work of Adams and McKellar and, interestingly, remarked:

"From the intensity ratio of the lines with K = 0 and K = 1 a rotational temperature of 2.3 kelvin follows, which has of course only a very restricted meaning." This material is easily accessed in Kragh's *Cosmology and Controversy* (p. 135). Yet another observation was made in 1962 by an American astronomer named William Rose, then at the Naval Research Laboratory, using an early radio antenna. He estimated a background of 2.5 to 3 kelvin but recorded his result only in his notebook. None of these efforts seems to reflect the painstaking removal of various extraneous effects from the observations that characterized the work of Penzias and Wilson, as did their faith in the validity of their measurements.

A recent measurement of the CMBR is, in our view, particularly interesting. Antoinette Songaila and colleagues measured the ambient temperature in a gas cloud near Zeta Ophiuchi by determining the population of excited states in a gas molecule—namely, CN. The relative number of molecules in two states led to an inferred temperature of 8 kelvin for the radiation suffusing the cloud. There have been similar measurements nearby, verifying the CMBR temperature at 2.725, but this cloud was at a significant distance cosmologically, at a scale-factor distance of $L = 2.9$ (nearby $L = 1$). The value 2.9 agrees nicely with a calculation of 8/2.725, which is what one would expect the CMBR to have been in the past and at a distance of $L = 2.9$; according to the properties of Planck radiation expanding adiabatically, the temperature T should vary inversely as the scale factor: $T/T_0 = 1/L$. Songaila's rather neat result thus lends credence to the interpretation of the CMBR.

It is too bad that just about everyone involved in the early history of the CMBR overlooked the earlier observation and analysis by Adams and McKellar of a background temperature in terms of the rotational temperature of CN as seen in space in absorption. The opportunity was missed by all concerned, including the authors of this book.

We should comment further on the question of whether at our urging the CMBR could have been detected earlier than the observation of Penzias and Wilson, and whether we pursued the possibility of its detection prior to 1965. We note again that the radiation had been detected in the early 1940s by Adams and McKellar, but neither we nor anyone else made the connection. After our 1948 prediction we presented talks at the Naval Research Laboratory and at the National Bureau of Standards in which we discussed the possibility of detection, but the radio astronomers with whom we spoke thought the radiation was not detectable with the technology of the time. We also gave invited papers on the subject to two different meetings of the American Physical Society. For one of these the society issued a press release that was picked up nationally, describing in some detail our prediction of a residual radiation from the Big Bang (couched for some obscure reason in a different temperature scale); no observers picked up on the suggestion. For the first several years after our 1948 prediction we were also somewhat inhibited by Gamow's insistence that the prediction was not real (as was also suggested by Edward Teller) or, that if the radiation did in fact exist, then it would be difficult to sort out the CMBR from other sources giving equivalent energy densities in the vicinity of the Earth, such sources as particulate cosmic rays and integrated starlight. Our colleague James W. Follin Jr. discussed the observational problem as late as 1954 with astronomer Allan Sandage at Mount Wilson–Palomar, suggesting that one might look for it in a rocket experiment. Writing to us in 1989, Sandage recalled this visit and stated that at that time he did not see any way to make the measurement.

In his widely read book, already referred to, Weinberg writes, "One may ask: When in fact did it become technologically possible to observe a 3-kelvin isotropic radiation background? It is difficult to be precise about this, but my experimental colleagues tell me that the observation could have been made long before 1965, probably in the mid-1950s and perhaps even in the mid-1940s" (p. 126). In retrospect, it appears that observations were indeed made but that their meaning was missed.

So we would argue that between 1948 and 1954 we did make a number of attempts to get some observers interested in looking for the radiation. If Weinberg is right, we were talking to the wrong people. Perhaps, as he notes "The important point is that the radio astronomers did not know that they ought to try" (p. 127). In the final analysis it is probably also the case that most of the scientific community did not take the Big Bang model seriously until the added evidence of the observation by Penzias and Wilson came along. Weinberg's book was probably itself an important turning point in acceptance of the model by the scientific world of physics and astronomy.

Gamow was not aware that we had asked a number of observers about looking for the radiation, and on other grounds he argued with us that if it did exist, it must be unobservable. We reiterate that he had come around by about 1950; in a semipopular article on cosmology in the magazine *Physics Today*, he referred to the existence of a 3 kelvin background radiation as a given attribute of the Big Bang model, though no reference whatever to its prediction or where he got the number. In an interesting historical footnote, Joseph Weber, until 1949 a naval officer and later well known for his attempt to detect gravitational waves while at the University of Maryland, reports that he was looking for a project to do in pursuit of a Ph.D. at Maryland, where he had already been given a professorship. Among those to whom he spoke was Gamow; he discussed using his skills in microwave engineering and was dissuaded, although presumably Gamow was aware of our prediction at that time. We suggest that this occurred during the 1948–50 period when Gamow knew about our result but either did not believe it or did not believe the CMBR could be observed.

Alpher gave a colloquium at Rutgers University (proximate to Bell Telephone Laboratories) after the announcement of the Nobel award to Penzias and Wilson in 1978 and before the award ceremony. Neither Penzias nor Wilson attended the colloquium but Penzias contacted Alpher, who then visited Penzias at his home. It was during an intensive period stretching over more than a day that Alpher told Penzias, who at that time was not really versed in cosmology, about our work on nucleosynthesis, our prediction of the temperature of the CMBR, and much of the technical background in the entire subject area.

In introducing Penzias and Wilson for the Nobel Prize at the December 1978 ceremony, Lamek Hulthen of the Swedish Royal Academy of Sciences pointed out that the explanation of the Penzias-Wilson observation by Dicke, Peebles, Roll, and Wilkinson "leans on a cosmological theory developed about 30 years ago by the Russian born physicist George Gamow and his collaborators Alpher and Herman. Starting from the fact that the universe is now expanding uniformly, they concluded that it must have been very compact about 15 billion years ago and ventured to assume that the universe was born in a huge explosion—the 'Big Bang.' The temperature must have been fabulous: 10 billion degrees, perhaps more. At such temperatures, lighter chemical elements can be formed from existing elementary particles, and a tremendous amount of radiation of all wavelengths is released. In the ensuing expansion of the universe, the temperature of the radiation rapidly goes down. Alpher and Herman estimated that this radiation would still be left with a temperature around 5 kelvin. At that time, however, it was considered out of the question that such a radiation would ever be possible to observe. For this and other reasons the predictions were forgotten" (Les Prix Nobel [Stockholm: Almqvist and Wiksel, 1978], p. 22).

Although our good friend and colleague Gamow did not at first believe that our prediction of 5 kelvin was meaningful, useful, or amenable to observation, and several years elapsed before he took it seriously, thereafter he wrote about the subject in a number of papers. These suffered on two counts: first, he rarely if ever referred to our first calculation; second, he redid the calculation in a way that was not correct and therefore led to an overestimate for the present radiation background temperature. Many authors, having finally become aware of the early work of Gamow, Alpher, and Herman, nevertheless proceeded to attribute the calculation to Gamow, usually quoting one of his overestimated values. This we feel is an example of the Matthew effect, well known to historians and sociologists of science: the tendency of scientists to attribute a piece of work with which several names are associated to the author whose name is a household word in the world of science. Thus one finds in the literature many references to the prediction of the background radiation with attribution to Gamow et al., Gamow and colleagues, Gamow and students, or simply Gamow, with only a footnote actually referring to publications by Alpher and Herman.

Not infrequently, the idea of a CMBR is apparently attributed to Gamow alone on the basis of papers he wrote in 1946 and 1948 in which he argued that the nonequilibrium synthesis of the elements required a hot, dense beginning for the universe. We argue that on this basis alone the conventional Big Bang model might well be called the Gamow model. But let it be noted that in neither of these papers, which are often cited as the first to deal with a present CMBR, did Gamow even introduce this notion. One would have to read into the papers the inference that if the universe was hot and dense at one time, then there should be lower-temperature relic radiation now. A careful reading of Gamow's papers prior to the early 1950s shows no such inference.

We ran into another problem because some authors, looking at one of our later papers in which we showed a calculation resulting in an overestimate of the background temperature, criticized the result in a manner that cast suspicion on the basic model. Our overestimate of 28 kelvin, however, published in 1951, was predicated on a short-lived observation by the German astronomer Albert Behr that for several years was widely held as correct and which gave a rather high value for the present matter density in the universe: namely, 10^{-29} g/cm^3. As already mentioned, matter density is a parameter that enters the calculation of the present background temperature in the standard model of the Big Bang. We quickly corrected our calculations in the technical literature when it appeared that Behr's results could not be duplicated.

We feel moved to remark on the problems of scholarship for many scientists. By way of example, in 1949 we were invited to write a review of the origin of the elements and cosmological implications. We carefully searched the literature for anything germane and actually read and pretty well understood the more than 200 papers in our bibliography. We found errors of various kinds in a number of papers and corresponded with the authors to enable a correct presentation of their results in our review. We have already mentioned Hoyle's lack of response to a letter we sent him. We do have at least one other bone to pick with Hoyle. Perhaps annoyed by the fact that our 5 kelvin proposal was said to be quite close to the observed 2.725 kelvin, he suggested that it would be more nearly correct to consider energy densities rather than temperature; we have on occasion wondered if Hoyle made these remarks tongue-in-cheek. Since energy density varies as the fourth power of the temperature, our prediction was on that basis in error by a fraction $(5/2.725)^4$—rather larger than a factor of less than 2.

The rewards for scientific work are hardly venal. One does science for two reasons: for the thrill of understanding or measuring something for the first time and, having done so, for at least the recognition if not approbation of one's peers. Some colleagues argue that the progress of science is all that matters and that it is of little consequence who does what. Yet we cannot help noticing that these same colleagues are nevertheless pleased with recognition of their work and accept with pleasure and alacrity such approbation as election to prestigious scientific academies. Recognition and approbation not infrequently make getting research grants easier and may help in receiving promotions. But we should not indulge in sermonizing about the nature of science. On to more about the CMBR and the COBE satellite.

In November 1989 a satellite called the Cosmic Background Explorer was launched by a multistage Delta rocket into a polar orbit from Vandenburg Air Force Base in California; it is still in orbit but was turned off for science after about four years. (It is worthy of note that the launch was delayed a bit when the satellite had to be redesigned, in part because it was originally scheduled to be launched from a shuttle; that plan was scrubbed after the *Challenger* disaster of January 28, 1986, and the satellite was modified for a rocket launch.)Nearly 2,000 scientists and engineers were involved in its conception, design, implementation, and launch as well as the years of data taking.

The satellite was placed in a polar orbit, an orbit that circumnavigated the earth from pole to pole, and it tracked the terminator, the line between light and darkness from solar illumination. The instruments were so located in the satellite as to be shielded and out of the direct glare of the Sun. Data from the flight continue to be analyzed, although many very exciting results are already in the public domain. The Project Scientist was John Mather of NASA-Goddard, who was also responsible for FIRAS (far infrared absolute spectrophotometer), one of the instruments aboard, whose main purpose was to study the CMBR; Michael Hauser, also of NASA-Goddard, headed the work on another major instrument package called DIRBE (diffuse infrared background experiment), which searched for isotropic cosmic infrared background radiation in the spectral range 1.25 to 240 micrometers; the third instrument package, DIR (differential microwave radiometer), was developed by a team headed by George Smoot of the University of California–Berkeley to seek departures from isotropy in the CMBR. The principal purpose of the satellite was clearly to make measurements of cosmic electromagnetic background radiation. The story of the satellite and of the people involved in its enormous success is well documented in at least two current books, cited earlier: one by John Mather and John Boslough (1996); the other by George Smoot and Keay Davidson (1993). They are both excellent reading, and make it quite unnecessary for us to give more than a summary.

All the results from COBE were exciting and reflect the forethought as well as the attention to detail in the design and implementation of the project. Consider first the verification of the nature of the CMBR: COBE showed the spectrum to be blackbody, and to reflect a homogeneous isotropic background at 2.725 ± 0.002 kelvin, an astounding precision (see figure 5.2), probably the most precise Planck spectrum ever measured anywhere, be it in space or in the laboratory. Moreover, this observation is now regarded as one of the major pillars of the Big Bang model. Literally millions of data points were required to map the CMBR sky with the desired precision. The data were broadcast to a ground station at Wallops Island, Virginia, on daily passes of the satellite over the station and then processed at the NASA Goddard Space Flight Center, Greenbelt, Maryland, by Mather and the various instrument teams.

It had been strongly suspected previously from a variety of largely ground-based observations, as well as early observations made by a telescope aboard a high-altitude NASA aircraft (a program directed by Smoot), that there should be an underlying anisotropy or nonuniformity of this radiation, because any observing platform, the solar system, the Milky Way, and the Local Group (more than 20 galaxies) have an underlying motion at some hundreds of kilometers per second with respect to the last opaque distribution of matter from which came the CMBR. This movement amounting to an absolute frame of reference was well verified by COBE. When it was first studied from the NASA high-altitude flight program, the effect was dubbed "the great cosine curve in the sky," since the amplitude of the effect varied across the sky as a cosine function of the angle of observation. When the background was measured in the direction of motion, a bit of warming was observed, with an increase in background temperature of about 0.3 millikelvin, while in the other direction, where the measuring instrument was receding from the direction of motion, there was a decrease of the same magnitude. This dipole component of the CMBR measured aboard COBE is illustrated in figure 5.3. To obtain this pure result it was necessary to subtract out the CMBR background and other effects known to be extraneous primarily from the Milky Way.

A second and fascinating result from COBE was the observation of anisotropy of the background radiation on a scale of 10 parts per million. This anisotropy persists in the data when one subtracts out the dipole contribution due to motion with respect to the surface of last scattering (decoupling) and extraneous emissions from the Milky Way, as well as the basic CMBR. A whole sky map of this anisotropy is shown in figure 5.4. It presumably had its origin in fluctuations of temperature (and density) which were present when the CMBR broke away from its previously trapped state and began to propagate freely in the cosmos. The origin of this anisotropy would have been in the very early stages of the universe, long before decoupling occurred, and the anisotropies may represent either a fundamental set of fluctuations occurring during an inflationary phase or standing acoustic waves traversing the cosmos before decoupling. One of the convincing arguments for the reality of this anisotropy is that it was observed, basically without change, at

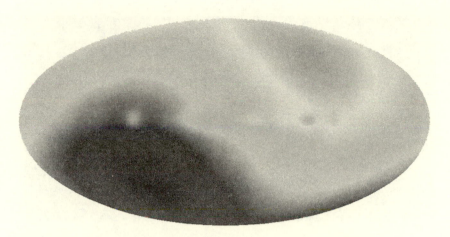

Figure 5.3. The dipole anisotropy as measured by the COBE satellite at an observing frequency of 53 gigahertz (10^9 cycles per second). $\Delta T = 3.353$ millikelvin. The anisotropy is shown as a whole sky plot. What is seen is the net effect of the motion of the observing platform (COBE) with respect to the cosmic microwave background as the radiations emerged from the material background in which it had been trapped, about 300,000–500,000 years after the Big Bang, and constitutes a CMBR frame of reference. The term *net* means the sum total of motion of COBE, Earth, the solar system, the Milky Way, and the local galaxy group, with the resultant velocity vector pointing toward the Virgo cluster of galaxies. In this figure the top right represents a blueshift (i.e., motion toward the CMBR frame of reference); the lower left shows a redshift (i.e., motion away from the CMBR frame) with a dipole amplitude, zero to peak, of 3.353 millikelvin, with the background radiation of 2.725 kelvin and various contributions from discrete sources in the Milky Way subtracted out. (Courtesy of NASA/GSFC and the COBE Science Team.)

several rather different frequencies, which argues against its being some kind of artifact.

We should point out again that there had been concern among cosmologists that the distribution of the cosmic microwave background across the sky, other than the dipole anisotropy due to motion with respect to the surface of last scattering (decoupling), was so uniform that in the absence of any other anisotropy there seemed no way to understand the subsequent formation of structure in the cosmos. They hoped ultimately to find deviations from uniformity that might provide seeds for the nucleation of galaxies and stars. COBE measurements have provided such seeds, but this is not the end of the concern, unfortunately, for the scale of the departures from uniformity is not consistent yet with any widely accepted theoretical calculation for the growth of structure.

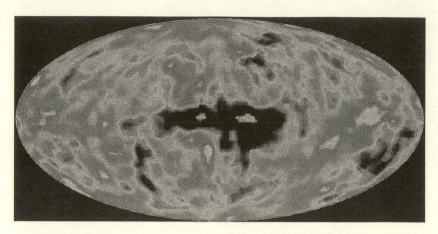

Figure 5.4. Primordial anisotropy: an all-sky map with the plane of the Milky Way running horizontally through it, as measured from COBE with four years of data. The monopole background (the 2.725 kelvin CMBR), dipole anisotropy, and any extraneous contributions such as Milky Way radiation sources have been subtracted out. What remains are departures from uniformity, with a scale ranging from −100 microkelvin (black to dark gray) to +100 microkelvin (white to light gray). The root mean square (rms) net anisotropy is 30 microkelvin. The angular resolution is approximately 10 degrees. These irregularities were presumably generated during the very early phases of the universe and are now observable—having survived through the period of decoupling of matter and radiation—and are variously interpreted as arising from cosmic strings, topological defects, quantum fluctuations, or compression (acoustic) waves prior to decoupling. The fluctuations are thought to be the needed basis for the subsequent development of structure in the cosmos. Recent observations, not shown, at about 1/4 degree resolution suggest fluctuations about three times higher than what is shown. (Supplied by NASA/GSFC and the COBE Science Team.)

Ongoing ground-based observations in considerable number are examining the anisotropies at other wavelengths than those observed with the COBE instrument. It is anticipated that there will be found a distribution in the scale of anisotropy with wavelength such as to enable a choice among theories for the growth of structure and, as well, as to give insight into a theoretical basis for the nature of the origin of the anisotropies in the very early universe. One cause may be the quantum fluctuations thought to arise from the applicability of the Heisenberg uncertainty principle (see the appendix) over a time scale of the order of the time constant for inflation, say 10^{-30} seconds. These fluctuations would then presumably have grown in size with inflation and also participated in the subsequent expansion of the cosmos up to the time of radiation-matter decoupling.

Figure 5.5 is a plot that summarizes the measurements through 1999 of the relative power in anisotropy features as a function of the opening angle of the radio frequency beams used to probe the anisotropies. It is not easy to visualize why the opening angle of the beams being used should matter. Consider that a radio frequency beam leaves an antenna as a cone of vertex angle equal to beam opening, and further consider that this cone intersects the surface of last scattering, whence the radiation comes. The beam sums up signals within its footprint, and since the nature of this summing up depends on the relative size of the fluctuations and the beam footprint, the summing would be expected to vary with the beam angle. One early option for understanding the power spectrum of the anisotropies was that the spectrum of the fluctuations was scale-invariant, in which case the summing up would not depend on the size of the beam footprint. This does not seem to be the case. There are of course other options, not yet clear from the observational data. The most recent view, however, may be summarized as follows: the anisotropies seem to peak at about just less than a 1-degree beam angle, which may lead to an interesting conclusion. This beam angle corresponds to a projected length on the surface of last scattering of about 300,000 light years, and we note that 300,000 years is about the cosmic age at the time of decoupling. Conclusion? Perhaps the anistropy scale reflects acoustic waves, with the longest and least attenuated wave corresponding to the size of the cosmos at that time.

Some Other Effects of the Cosmic Background Radiation

The ubiquitous CMBR has a number of effects that are of considerable scientific interest. We mention two. One is the already mentioned Sunyaev-Z'eldovich effect, named after the two Russian cosmologists who proposed its existence. In the nearly empty space in a cluster of galaxies, a very tenuous gas of nuclei and electrons is kept in an ionized state at an elevated temperature by radiation from the surrounding galaxies. When photons of the 2.725 kelvin background traverse this gas, they interact with the electrons present (which are very tenuous and

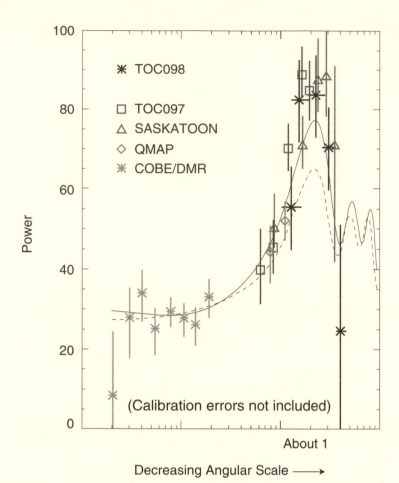

Figure 5.5. More on anisotropy. The vertical scale shows the power measured by a microwave frequency antenna with a given angular aperture, and is proportional to the average corrected departure of temperature within the beam from that of the uniform CMBR. The horizontal scale represents the beam angle of the observing antenna. COBE data on the left are at about a 7-degree beam angle. The TOCO data were taken at a microwave anisotropy telescope in Chile; analysis of these data is not yet complete. A joint U.S.-Italian project called BOOMERANG flew a telescope around the South Pole on a balloon for 10 days early in 1999 and measured anisotropies on a scale some 35 times finer than had been done previously. Other probes are scheduled in 2000 and later. BOOMERANG data suggest not only a peak at about about a 1 degree scale but subsidiary peaks at even finer scales, very suggestive all together of a spectrum of acoustic wave resonances, with the fundamental wavelength corresponding to the size of the visible universe at the time of breakout of the CMBR: namely, about 300,000–500,000 years after the Big Bang. (Previously unpublished figure reproduced courtesy of Lyman Page of Princeton University and Mark Devlin of the University of Pennsylvania.)

at a considerably higher temperature) and are scattered to a higher temperature; hence, observers see a hot spot in the background when looking at the gas in a cluster. This effect is now well documented and is being used to measure distances of galactic clusters. It has not yet achieved the precision of other measures.

A second effect is a cutoff in the energy spectrum of cosmic rays. We observe particulate radiation from outer space, mostly protons. Such radiation was first extensively studied by the late Victor F. Hess, then at Fordham. For a long time there was thought to be an upper limit to the energy carried by cosmic radiation; this had been suggested by K. Greisen of Cornell and G. T. Zatsepin and V. A. Kuzmin of the former Soviet Union and explained as arising from interaction of the particulate cosmic radiation with the ubiquitous cosmic microwave background photons. The expected cutoff was very high, but now observers have found some cosmic rays at energies greater than the expected limit; an explanation is not yet available.

6

Inflation
and the Very
Early Universe

Most of what we have reported up to this point is material with which we have been involved. Before proceeding, therefore, we must remark that much of what we say about an inflation era in the early universe is derivative; that is, we have not personally studied Big Bang cosmology at times which are very small compared with a microsecond but have followed what other researchers say about this early history.

Inflation theory is an active area of research, going in several directions. Some of these directions are in themselves intriguing, and inflation theories do address some of the basic conundrums of the standard Big Bang model; we deal with these below. In the early days of our work in fleshing out the Big Bang, we were (and remain) intrigued by the question of how to deal with what the general relativistic model indicated about the singularity it predicted, where, at all coordinate positions, the time zero was characterized by infinite temperature and density. Such a description is at odds with the intuition of scientists who work with cosmological models. Einstein himself said that at times very close to zero the general relativistic model should be expected to fail, technically, because the radius of curvature of the universe might be expected to be exceedingly small. With the introduction of inflation algorithms, it has become accepted that one deals with times as small as the Planck time, namely, 10^{-43} seconds. More of this in chapter 9 and the appendix.

In 1953 with James W. Follin Jr., our colleague at the JHU/APL, we undertook a detailed study of what was going on after the universe was about a second old, a time well before the onset of nucleosynthesis, which began some several minutes later. This epoch in the Big Bang does not require any new physics; high temperatures and the phenomena associated with them are accessible to scientists in laboratory-scale experiments, particularly when supplemented with well-founded and well-verified theories.

Several recent books discussing inflationary models are referred to in our recommended reading list; note in particular *The Inflationary Universe* (1997) by Alan Guth. He is regarded as one of the originators of the concept of inflation in the early universe, although there have been a number of alternative inflation models since his first work, including his own modifications. Moreover, the magazines *Scientific American*, *Discover*, and others have had articles on inflationary modeling from time to time, particularly since Edward Tryon suggested the universe began as a vacuum fluctuation and that it "is simply one of those things which happen from time to time" (*Nature* 246 [Dec. 14, 1973], p. 397) and Alan Guth of the Massachusetts Institute of Technology referred in *The Inflationary Universe* (1997) to the creation of the Big Bang universe as "a free lunch" (p. 15). Guth even supposed that given the right technology one could "create a cosmos in the laboratory" (p. 253). Other major contributors to inflationary paradigms include Andreas Albrecht, Andre Linde, Paul Steinhardt, and A. Vilenkin. Other theories too bear on the very early universe, such as string theories, where the names of Edward Witten and Brian Greene are noteworthy, and supersymmetry, but these are not central to the topics of this book.

Briefly and superficially, each of the inflationary models of the early universe (and again we emphasize that there are several) supposes that at a time of the order of 10^{-43} seconds after the Big Bang (the so-called Planck time, discussed below, which is unarguably early), the content of the universe underwent a very rapid inflation in its dimensions, by as much as a factor of about 10^{+50} (which is unarguably large). This process may well have been finished by a time of the order of 10^{-30} or 10^{-35} seconds (still unarguably early). Certainly by the end of a very small fraction of a second after the Big Bang event, however described, the physical phenomena associated with the period of inflation would have ceased, and the expansion and cooling of the universe would then have continued as described by the standard Big Bang model. The speed with which the inflation occurred would not violate the relativistic speed limit on the velocity of light, because there would have been no information transfer, as is required for this limit to be meaningful.

We emphasize that various versions of inflationary models do not conflict with or replace the Big Bang model (as has too often been erroneously stated) but, rather, are complementary. The former can make a smooth transition into the latter at times which are very short, on a scale of microseconds. Moreover, although the paradigm of inflation is by no means as firmly founded in observation as the rest of the Big Bang model is today, it does yield at least qualitative or better than hand-waving arguments about some of the outstanding puzzles of the Big Bang cosmic scenario.

The Need for "Inflation" and Other Fixes?

What are the conundrums which, without inflation or some other fundamental modification of the Big Bang model at very early times, are puzzling? One is called the *horizon problem*, which, simply stated, is this. The data from the COBE satellite showed that the universe is suffused with blackbody photons with a temperature of 2.725 kelvin, and that this temperature is very nearly the same when measured in any direction (though there is a small deviation with a total difference or amplitude of the order of three millikelvin in opposite directions). This result indicates a motion of the observer, taken together with the motion of Earth, Sun, Milky Way, and our local cluster of galaxies, toward a point centered with respect to the surface from which what is currently observed as the microwave background radiation emerged in the past from the opaque plasma and began to travel freely through the universe. This emergence defines a horizon. But if we look at any two diametrically opposed points in the sky, the background radiation temperature seen there is the same, and yet the radiation emanated from a time early in the universe such that these opposed points—opposite horizons, if you will—would be separated by very nearly twice the age of the universe and could not have been in contact earlier on. How then did the whole universe come to be at the same temperature with such delicate precision, and therefore exhibit such exquisite large-scale uniformity, when parts of it could not have had causal contact with other parts during the lifetime of the Big Bang?

There are of course smaller-scale departures from homogeneity in the form of stars, galaxies, and galactic clusters, but these were undoubtedly formed after the essential homogeneity exhibited by the cosmic microwave radiation background had been established and was emanating from the surface of last scattering horizon. In addition, there are the already mentioned extremely small departures from uniformity in the cosmic microwave background radiation (on a scale of 10 parts per million and different from those that result from motion of the observing platform). These, on the one hand, appear to have developed from extremely small fluctuations in temperature or density during an inflationary era, if there was one, and, on the other hand, to have enlarged as the universe expanded and provided "seeds" for the development of the structures we now observe in the universe, albeit by mechanisms that have yet to be satisfactorily sorted out by cosmological theorists. Recall that until the COBE data revealed these small fluctuations, there was genuine concern among theoretical cosmologists that the extreme uniformity of the cosmic microwave background made it very difficult to understand the formation of the structural details now exhibited by the universe.

Scientists are currently using special instruments at many telescope sites to determine the power spectrum: that is, the distribution of sizes and intensities of these tiny fluctuations across the sky. The original COBE observations were made with a satellite-borne instrument that integrated over the footprint of the observing beam, a cone with an opening of 7 degrees. The *footprint* is an interesting concept which bears repeating. The COBE instrument accepted radiation coming to it within a cone defined by the instrument design. Think of this cone as extending out to the surface of last scattering (i.e., the surface from which, in the expansion, the background radiation emerged, having essentially no further interaction with the matter in which it had been trapped until then) and intersecting it. If one visualizes that surface as a sphere covering the entire sky, centered at the observer, and extending 360 degrees, then the cone's intersection is a circular footprint, extending 7 degrees—a small fraction of the total extent of the sky. Therefore, we have a cone intersecting the surface of a sphere, the horizon for the blackbody radiation.

The devil may be in the details, for it is hoped that establishing the distribution of fluctuation sizes will lead to an understanding of the mechanisms responsible for forming these fluctuations in the very early inflating universe. Given the size and content of the observable universe just before and during inflation, it is most likely that the fluctuations are quantum in nature and may ultimately be understood in the context of the uncertainty principle of quantum mechanics. In fact, one of the underlying motivations for exploring an inflationary paradigm is to help in developing a theory of quantum gravity, since conditions during this inflation period must have been so extreme as to require the application of quantum mechanics—the physics of the small, or microcosmos, rather than the physics of the large, or macrocosmos. In other words a quantum mechanical theory is needed which subsumes not only the elementary particles and radiant energy of the early universe but also the extremely high gravitational fields that must have been extant and, as well, the possible formation of particulate matter, perhaps from the energy in the gravitational field. Proponents of inflationary paradigms argue that the homogeneity and isotropy of the standard Big Bang model are assumptions, and they are prone to develop theories that eschew these and other Big Bang assumptions. As previously pointed out, there has always been a concern about whether in the standard Big Bang model the classical general relativity that underlies it really is applicable at very early times; recall Einstein's concern about general relativity close to the singularity. Going back in time, as one approaches the singularity, one wonders about the increasing curvature of space, which in simple terms would become exceedingly small as zero time is approached. Proponents of the inflationary paradigm suggest that there is a fundamental limit to the radius of curvature of space: namely, the so-called Planck length, which is 10^{-33} centimeters.

As this is written, there seems to be a convergence of opinion that the basic scale of the inhomogeneities observed in measurements made since COBE is of the order of 1 degree. The situation is still fluid, and one should not yet discard on the basis of existing data that the number density of occurrences of departures from isotropy is scale-invariant, which would agree with a model for the very early universe developed by Edward Harrison and Yakov Zeldovich. But there are many ongoing observations, and there are no final answers, of course, as is the fate of science. The reader may want to revisit figure 5.5 to review the anisotropy data.

A second puzzle has to do with the fact that the Big Bang model requires a constant ratio (constant throughout the duration of the expansion) of the number of cosmic blackbody photons to the number of baryons, a dimensionless ratio with a value of 10^9 to 10^{10}. This number has several interpretations; one is that it expresses the entropy per baryon in the cosmos, entropy being a measure of disorder. The numerical value of this ratio is assumed in the usual Big Bang theory, since it is completely equivalent to a statement of the density of matter prevailing in the universe during the initial minutes when light nuclei were formed by thermonuclear reactions, and, as stated earlier, this has been the one free parameter in the Big Bang model. It would be better from a theorist's viewpoint if this number could be calculated or developed from first principles rather than having to be assumed. Moreover, the numerical value of the ratio, however it is interpreted, is both very large and unchanging in the framework of the standard Big Bang model. Qualitatively, at least, several inflationary paradigms and some unfortunately obscure work of Ilya Prigogine appear to provide one rationalization for this very large ratio (see I. Prigogine, *International Journal of Theoretical Physics* 28 [1989]: 927). Prigogine's ideas are perhaps simplest to understand, for he points out that the conversion of negative gravitational potential energy into particles during inflation automatically increases entropy, given that entropy is directly related to the number of particles (and therefore the number of different states into which energy feeds) extant in the expanding medium. Creating more particles also creates more entropy. There are other sources of entropy change as the universe evolves, such as the disorder introduced by the formation of stars and galaxies, but their contribution is orders of magnitude too small to affect the large entropy per baryon from the early universe.

A third puzzle has to do with the ratio of the total density in the universe to the critical density, the ratio that we call Ω (omega). This density divides the solution space of the Big Bang model (i.e., the regime of all possible solutions of the basic equations of the model, as seen, for example, on a graph of the scale factor of the expansion plotted against the elapsed time from the Big Bang event, illustrated in figure 6.1; see also the "divine curves" in the appendix) between an open, ever expanding model, on the one hand, which becomes asymptotic to a constant expansion rate, and, on the other hand, a closed model that will ultimately slow down, stop expanding, and then collapse toward what has been called the Big Crunch. This collapse may be simply a phase in oscillatory behavior, with the universe going through a series of expansions and collapses. If the universe is flat, the scale factor approachs a constant value, with expansion rate equal to zero. These days one can certainly argue on observational grounds, without being challenged seriously, that this ratio must lie between a few tenths and a number somewhat bigger than unity. Were it to lie outside this range, it would not be easy to understand the structure and evolution of the cosmos. It is in fact viewed as

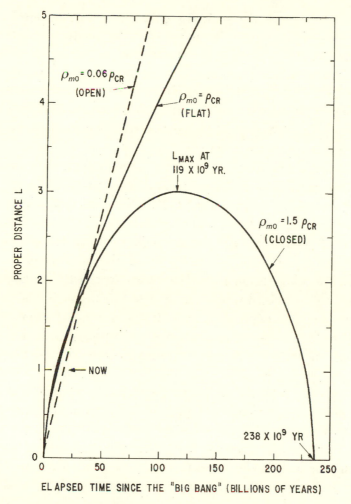

Figure 6.1. The time dependence of proper distance (the scale factor in the expansion) for the cases of an open or low-density universe; a flat model with Newtonian behaviour (no curvature in space-time), whose mean density is equal to the critical density which separates open from closed models; and a closed model, in which the mean density is greater than the critical density and which attains a maximum extension, when the expansion rate goes to zero and then becomes negative, so that the model collapses back toward an initial state, called the Big Crunch by many. The present age is very close to the origin in this figure, where $L = 1$ is arbitrarily taken as the present value of the scale factor. This shows why it is difficult now to sort out which of the models is the best one to use. (Reprinted with permission from *Proceedings of the American Philosophical Society* 119 [1975]: 325–48.

being so close to unity that some argue that there must be some kind of natural law requiring it to be precisely unity, both early in the universe and now. If there was a period of inflation with a supposed expansion of the dimensions by a factor of about 10^{+50}, any departure from an initial value of unity in Ω could have been driven very far from unity. Proponents of inflation therefore insist that whatever physical processes were operating during inflation must have been such as to keep this ratio invariant and precisely equal to unity.

We understand this argument but are not yet completely persuaded. An idea not particularly new or original with us but reasonably obscure in the literature is that if one takes the mathematical description of the standard Big Bang model back to very early times, one finds that the deviations from unity for Ω are exceedingly small also, without invoking inflation, being of the order of 10^{-16} times the age of the universe in seconds. This translates to a deviation of one part in 10^{59} at the Planck time of 10^{-43} seconds, which is indeed close to unity, and it is still extremely close to unity even as late as 300,000 to 500,000 years into the expansion of the universe, when the expansion changed from being controlled by the density of radiation to being controlled by the density of matter. To put it somewhat differently, we do not find any a priori or compelling reason for invoking inflation simply to maintain a density ratio equal to unity throughout cosmic history. It is therefore tempting to argue that having Ω be exactly unity for all time is a matter of predilection, rather than a model necessity, and may be no less *ad hoc* than other assumptions for the model Ω. (The early-time approximation for the density ratio Ω in a radiation-controlled universe is given in the appendix.)

Let us make some additional remarks about the cosmological constant Λ. There may be several contributions to the value of Λ, particularly if one takes into account recent evidence that there was an acceleration of the rate of expansion in the earlier universe (as observed in the velocity of expansion of structure in objects seen when the universe was much younger). If so, then there must have been a driving force, and cosmologists view this force as deriving from a negative gravitational potential, which can in turn be attributed to an added term in the general relativistic equation describing the model. This added term involves the cosmological constant Λ, a situation we discussed earlier. The energy density associated with it would contribute to the value of Ω, and we can denote it as Ω_Λ. As already mentioned, there is some amount, not yet all that well determined, of dark matter, which contributes to the energy density in the cosmos. We can call this Ω_m. Finally, there is a contribution of baryons, determined by what is required in the early universe to explain light-element nucleosynthesis. Call this Ω_b. If we go with the argument that Ω (total) should be unity, at least early and perhaps for all time, then

$$\Omega_b + \Omega_\Lambda + \Omega_m = 1$$

Another concern addressed by the inflationary paradigms is the possible large-scale presence of free magnetic charges or monopoles, as predicted by the Dirac theory of the electron. Free magnetic charges are for the present a theoretical construct. Anyone who has had occasion to work with bar magnets, with a north pole on one end and a south pole on the other, will recall that if one breaks a magnet in two, the fragments will each have a north and a south pole. The monopole, if it exists, is either a north or a south pole, with no opposite pole to give nearby closure to the magnetic field lines usually associated with a magnet. Since these matters are beyond the scope of this book, suffice it to say that the universe does not appear to contain large numbers of monopoles, and some recent theoretical studies suggest that primordial monopoles do not in fact exist. Despite strenuous efforts by experimentalists, at most a single monopole event has been observed—in an experiment that has never been reproduced and may well have been an observational artifact.

Inflationary paradigms have also at least pointed the way to an understanding of why the universe appears to be composed of one form of matter, rather than equal amounts of matter and antimatter. Recall that every elementary particle has an associated antiparticle. Thus, there are electrons, which are negatively charged particles of a given mass and seemingly of zero size, but there are also positrons, which are just like electrons except that the charges are positive. We frequently refer to baryons, particles whose masses are of the order of the mass of the protons, and antibaryons, the associated antiparticles. More than 40 years ago we suggested that the matter-antimatter abundance ratio would have to come out of a knowledge of the physics of the early universe, and that still seems to be the case. As with monopoles, there is as yet no final answer. There is, however, at least observational evidence that in some situations the rate of certain reactions in which matter transforms into antimatter do not behave symmetrically for the reverse reaction, the transformation of antimatter into matter. A nuclear transformation which exhibits symmetry-breaking in the decay of the K^+ meson, one of the zoo of elementary particles, which was observed by Val Fitch and James Cronin to have two alternative decay schemes. K^+ can decay into pions K^0 or a K^1 with the fact that there are two decay paths constituting breaking symmetry. In this case the alternative decay products have different parities (left- or right-handed spin), so parity is violated.

We have by no means exhausted the list of problems associated with the early universe. The question most asked, we suspect, is what came before the Big Bang. As a corollary, some ask which came first, the universe, or the laws of nature. Let us have a brief stab at these. The conventional Big Bang model suggests that the expansion started everywhere in the universe (i.e., all three spatial dimensions) at time zero. This is where the cosmos began, and we might argue that asking what went before something that "began" is a bit like asking about latitudes 91 degrees north and 91 degrees south. Some people, for whatever reason, retreat to the idea of an extramundane entity that started things off, but this does not seem to buy us any understanding; in effect, it obviates any need for understanding, although it makes some people feel more comfortable. Alternatively (and there are inflation gurus who push this idea), the universe began with a physical "nothing," a vacuum. In a vacuum, we know observationally, pairs of particles and antiparticles can appear and disappear for short times without violating conservation laws. The duration of appearance and the associated energy are governed by the uncertainty principle, whose relevance we have no a priori reason to discard for early cosmic times. This physical principle, introduced in the development of quantum mechanics by Werner Heisenberg and firmly founded in observation, states that the product of the fluctuation energy times the duration of the fluctuation numerically equals or exceeds a numerical constant: the Planck constant. The cosmos, then, may have derived from such fluctuations that did not decay away. Moreover, such fluctuations, enhanced in subsequent expansion, may be deeply involved in the parts-per-million deviations from uniformity in the cosmic microwave background radiation.

Another objection occasionally raised to the Big Bang is why the universe is as big as it seems to be, where big denotes the dimension c (velocity of light) times the age of the universe, a product which is of the order of 10^{28} centimeters if the age is about 15 billion years. This is the distance to the observable horizon, in principle, and contains about 10^{80} baryons (a number still called the Eddington number, after the English astronomer who first suggested it). In one view of the early universe, all that existed was contained in a volume of radius c times the Planck time, or 10^{-33} centimeters, at the Planck density of 10^{95} grams per cubic centimeter, containing therefore about 20 times the Planck mass, or about 4×10^{-4} grams. It seems to us that this begs the issue of whether the universe was infinite or finite but unbounded, with more and more material being subsumed within the observable horizon as expansion and cooling went on. Moreover, since inflation began everywhere at some initial time, there could be an infinite number of universes that we cannot ever observe because they lie beyond the distance ct_{age}, which defines the age of our bit of reality. We might as well dismiss one often raised question about this situation. The observable region enlarges at the velocity of light, which is the speed limit for usable signals in the universe, according to relativity theory. So even if one can imagine a neighboring region of space-time whose observable horizon impinges on ours, the sum of the two rates, though naively identified as $2c$, cannot exceed c; hence, knowledge of this neighboring region is precluded. It is difficult to model seriously that which we can never hope to observe.

Yet one more question: whence came the matter and radiation that filled the universe after inflation and was the precursor of that which populates the observable universe—particularly since the Big Bang assumes conservation of baryons through the expansion, with radiation expanding adiabatically? Several approaches to an answer all depend basically on the idea that there was a *phase transition* during the inflation period, which led to the appearance of elementary particles and blackbody radiation. For those not familiar with phase transitions, the situation resembles what happens when, say, water is heated and changes to water vapor. There is a similarity between this view and the view of Prigogine, mentioned earlier. We could be wrong, but then it should be clear to the reader by now that we are not yet sold on the inflationary paradigm.

The supposition that the expansion began everywhere at time zero seems to us to avoid the criticism that the Big Bang model does not explain the synchronicity of the start of the expansion. One suggested alternative is that the universe is like a foam in structure, with each element of foam starting to expand at the same time. In the end, as we just explained, these little regions would not be visible, one from another, and one has a result without observable consequences. The foam could in fact be infinite in extent in the model, but to what end?

Another criticism, and it may be less a criticism than a philosophical question is why we see the universe we see. Is there some magic in describing the universe with just four dimensions, three of space and one of time? Why are there not more dimensions, observable or not? There are theories involving strings, the concept of supersymmetry, and so on which do not limit the number of dimensions. Another mystery is why there is such a range of masses among the elementary particles. Why did the cosmos develop with neutrons and protons having masses some 2,000 times those of electrons? Moreover, why does the baryon have a finite size, while the electron and its antiparticle appear to be point particles? The universe is yet filled with mystery.

Planck-Ginzburg Numbers

In the foregoing discussion we have alluded to times of cosmological interest which are approximately 10^{-43} or 10^{-35} seconds, times that are implicitly introduced as fundamental in the literature on inflation. What do these very short times mean? In what follows we refer in particular to units of mass, length, and time and to derived units of density, temperature, and so on, which are named after the German scientist Max Planck, with more recent work by the Russian Viktor M. Ginzburg. Readers are usually given short shrift as to where these units came from and why they may be fundamental in cosmology. The popular and technical literature frequently invokes a time called the Planck time—namely, 10^{-43} seconds—as though it were a fundamental division of time handed down from on high. Perhaps it is, and it may indeed be the fundamental quantum of time.

There are several ways to try to understand such numbers. One is contained in grand unified theories, called GUTs. But we believe that the philosophical underpinnings of the inflation model may be traced to the early work of Planck (early in the twentieth century) and more recent work of Ginzburg, which preceded GUTS. Let us consider a simple view.

What Planck proposed is that in physics one should use a set of units of mass, length, and time based on appropriate combinations of natural constants, c, h, and G. These are selected because c, the velocity of light, is basic in special relativity; h, the Planck constant, plays a basic role in quantum phenomena (and is the constant of proportionality between the energy of an electromagnetic wave and its frequency); and G is essential not only in Newtonian physical descriptions of gravity but also in the general theory of relativity. There are several ways of getting at these units. For example, one can derive them from a procedure known as dimensional analysis. Alternatively, one can derive them from the so-called grand unified theories, which suppose that at a sufficiently high temperature, all the four fundamental forces of nature have the same magnitude. These forces are gravitation, the so-called strong force (the force involved in the interaction of nuclear particles), the so-called weak force (involved in the radioactive decay of certain nuclei with the emission of a positron or an electron), and finally, the electromagnetic force (the force between charged particles). The magnitudes of these four forces, all of which depend on what are called coupling constants, are temperature- or energy-dependent and contain c, h, and G. The dependence is such that at very high energies (or temperatures), in some cases as given by the Planck units listed below, all the forces are thought to be equal in magnitude. There is theoretical and observational evidence that the weak force and the electromagnetic force come together at energies that can be reached by the largest particle accelerators used in physics laboratories. The strong force has not yet been explained in this sense in laboratory experiments. There is not yet evidence, moreover, for a unification of gravitation with the other fundamental forces.

A discussion of the Planck numbers is given in the appendix. The numbers are as follows, and we call your attention to the fact that most of them have surprising magnitudes, mostly quite counterintuitive, with little or no application in conventional physics:

Planck mass (m_P)	$m_p \approx 2 \times 10^{-5}$ grams
Planck length (l_P)	$l_p \approx 10^{-33}$ centimeters
Planck time (t_P)	$t_p \approx 10^{-43}$ seconds

From these we can derive the following quantities:

Planck density (ρ_P)	$\rho_p \approx m_p/(l_p)^3 \approx 10^{95}$ grams/cubic centimeter
Planck energy (E_p)	$Ep \approx (m_p \times c^2) \approx 10^{28}$ electron volts, or 10^{19} Gev
Planck temperature (T_p)	$T_p \approx E_p \neq k \approx 10^{32}$ kelvin

The Boltzmann constant, k, with units ergs per degree kelvin, is, for our purposes, a constant of proportionality between the energy of a system in thermodynamic equilibrium and its temperature (the energy of a system, E, is simply k times the temperature; with temperature in absolute kelvin degrees, the energy is in units of ergs).

The only one of these quantities which is not so large or small as to boggle the mind is m_p, and its physical significance is not clear. Its about the mass of a grain of sand. To put the other numbers in some kind of context, we note that the mass of a nucleon is of the order 10^{-24} grams, which is a very small fraction of m_p; that the density of the atomic nucleus is of the order 10^{14} to 10^{15} grams per cubic centimeter; that the temperature at one second after the Big Bang event in the standard model is calculated to be 1.5×10^{10} kelvin; and that the size of a typical nucleus is of the order of 10^{-13} centimeters. There is no obvious physical system, other than the hypothesized very early universe undergoing inflation, which appears to involve a time interval of 10^{-43} seconds. Moreover, the Planck length, which is much smaller than anything observable, is suggested by inflation protagonists as the minimum radius of curvature, which is meaningful in the space-time curvature of the medium in general relativity.

As an aside, let us consider the mean density of baryons presently observed in the universe, which is a very small number: about 10^{-30} grams per cubic centimeter. This number is obtained by adding the mass of galaxies and galaxy clusters in a given volume and dividing by the volume, thus smearing the density out as though it were a uniform gas. Dividing this mass density by the mass of the nucleon, about 10^{-24}, we get 10^{-6} nucleons per cubic centimeter (one nucleon per cubic meter). If now we multiply this last number by the volume of a cosmos with a radius of 15×10^9 light years (one light year is 9.5×10^{17} centimeters, and the number is the distance to the horizon which is observable in principle, assuming about 15×10^9 years as the age of the universe), we find about 10^{80} nucleons in the cosmos, a number which, as we have mentioned, had some currency in earlier cosmological literature as the Eddington number for the visible universe. The number is accorded some significance in dealing with the anthropic principle.

Rather than trying to deal with the several different approaches to inflation, let us try to sketch the principal elements that are common to all approaches. It is supposed that at some very early time after the Big Bang the universe exhibited an expansion that was for all practical purposes much more than exponential in nature. Linear dimensions in the cosmos increased by a factor of the order 10^{50} during a time from about

10^{-43} to about 10^{-30} seconds. This large expansion would have many effects. For one thing, it would so dilute the abundance of magnetic monopoles as to make their existence today moot. Another and most important consequence is that the expansion would have a profound effect on quantum fluctuations in the early universe, perhaps stretching them so that they constitute the density or temperature fluctuations that are now visible in the cosmic microwave background radiation and provide the basis, the so-called seeds, for the subsequent agglomeration of material into the structures we now see in the cosmos. We repeat that the mechanism for this agglomeration is very much a research question today.

We should expand upon several things here. Again we remind the reader that we have not been responsible for the development of the varioius inflationary paradigms. One version, borrowed from the writing of Andre Linde, suggests the ubiquitous presence of a *scalar field* in the universe. Scalar fields are characterized by the specification of a single variable: for example, a scalar field of temperature. A *vector field* has several variables for specification, including one for magnitude and several for direction: for example, a field of fluid flow has a direction and a magnitude of flow at each point. A scalar field, which is not observable unless there is a gradient, might exist when the four fundamental forces all have the same magnitude, which would be the case at an enormously high temperature such as the Planck temperature mentioned above. With the cooling of the cosmos, these forces would become separately identifiable, in a process known as symmetry-breaking. A characteristic of the scalar field is that it would diminish slowly with time in the expansion, much more slowly than the diminution associated with the density of energy in the early universe. Another feature of the scalar field would have been to make some kinds of particles heavy, such as the W and Z particles involved in the weak interaction, while rendering massless such other entities as photons. The energy in the scalar field would diminish slowly toward a minimum value, where its control of the expansion, which would have plenty of time to go on, would switch over to conventional expansion, having been exponential up to that time. Oscillations of the scalar field around the minimum would have produced the various early elementary particles. The exponential expansion would have started at the Planck time and finished at, say, 10^{-30} seconds. The expansion would have been so rapid that an initial volume containing only a few particles would end up with a particle density and number almost too large to consider as being in the visible universe.

Let us consider the concept of quantum fluctuations, which may have generated the departures from uniformity during the expansion phase. These fluctuations arise because of the more than likely applicability of the Heisenberg uncertainty principle. In what is effectively a vacuum in the very early universe, pairs of particles and their antiparticles can appear and disappear without affecting the local conservation of energy. The created particles have an associated energy of ΔE and can last for a time Δt, with the proviso that the product of these numbers be of the order of or larger than the Planck constant h. The fluctuations may also be simply some kind of energy or temperature fluctuation. In the rapid exponential expansion, these fluctuations would be stretched, and very likely some would survive to appear as departures from uniformity in the post-expansion phase. Aside from these tiny departures from uniformity, the expansion would have been sufficiently rapid to homogenize the temperature and density distribution of the universe as it entered the period of expansion governed by the standard Big Bang model.

What we now know about the cosmos does not demonstrate the validity of inflationary paradigms, but the qualitative solution of some of the mysteries of the standard Big Bang model renders study of such paradigms a fascinating area for further research.

7

Further Discussion
of Alternatives

With some hubris, we would argue that the standard Big Bang model is the best model of the observable cosmos that has yet been devised. Does it have problems? Yes! Might theoretical and observational developments in the future bear on the model's correctness at some level of detail? Very likely! Could any such developments rule out the Big Bang model as a descriptor? Perhaps. It is not at all impossible that something new in the way of theory or observation will send cosmologists back to the drawing board. Moreover, the Big Bang model did not appear out of nowhere, unopposed. Other attempts to understand the cosmos were being put forward at the same time, and their successes and failures certainly influenced the Big Bang model. There are still valiant efforts by some cosmologists to come up with alternatives, including in particular an argument that the correct model is a modified Steady State model. This quasi-Steady State model was presented, as we have already mentioned, in 1999 by G. Burbidge, F. Hoyle, and J. Narlikar, in *Nature* and in *Physics Today*, and a strong refutation was published in the same issue of *Physics Today*, by Andre Albrecht, otherwise known for his work on inflationary models.

Other Models

Let us list a few of the serious and major efforts of the recent past. These are brief summaries, since we have discussed some of them in one way or another in earlier parts of this book, or have made judgments about the importance of those alternatives that were being developed while we were more scientifically active. Our coverage therefore is far from exhaustive. Moreover, we have made value judgments about some of the proposed models on the ground that they appear to have little or no substance.

Equilibrium theories. Theoretical equilibrium studies have related primarily to questions of the origin of the chemical elements. The driving forces for such theories have been the realization, first, that observed relative abundances really are cosmic and that the formation of the elements should therefore be of cosmological interest; second, that there are mass defects in nuclei that can be dealt with as thermodynamic or Gibbs potentials, as expanded upon below. This makes calculations in an equilibrium situation relatively simple. Such theories have seen a lot of action, historically speaking, and indeed are still important in understanding the origin of heavier nuclear species in stellar interiors, though with less direct cosmological relevance than the formation of the lighter elements early in the expanding universe. Put simply, the present view of Big Bang adherents is that light elements (say, up to lithium) are formed in a cosmological milieu, the early expanding universe; the formation of others of the lighter elements, some lithium isotopes, beryllium, and boron, occurs later on with the breakup of nuclei by cosmic rays, whose energies are sufficient to split off chunks of nuclei; still heavier elements are formed in stellar interiors, later in the evolution of the universe.

For the sake of completeness, consider the basis of equilibrium theories. Every particular type of nucleus is characterized by a specific number of neutrons and protons. As we have mentioned earlier, the masses of the individual neutrons and protons add up to more than the total mass of the nucleus. The discrepancy is called a mass defect; the defect represents energy bound up in the forces holding the neutrons and protons together. From the usual relation due to Einstein, $M = E/c^2$, there is an equivalent mass associated with the energy of binding. Because the mass defect plays such an important role in equilibrium theories, we show a schematic plot of mass defects (expressed as binding energies per nucleon or baryon in the nucleus), plotted against the atomic weight, in figure 7.1. We note that the curve rises from low values at very low atomic weights to a peak in the vicinity of iron, then diminishes with increasing atomic weight above iron.

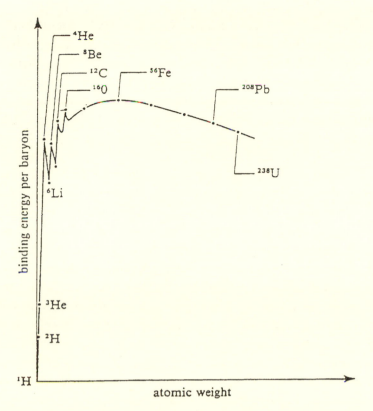

Figure 7.1. The mass defect (binding energy) per baryon (i.e., divided by the total number of neutrons and protons) of atomic nuclei, plotted as a function of atomic weight (the total of the number of baryons). Notice the high binding energy of the helium-4 nucleus and of iron, which tells something about the high primordial abundance of helium and the production of iron, nickel, and cobalt in the last stages of fusion reactions in stars exhausting their nuclear fuel. Helium-4 has about 7/8 of the binding energy per baryon of iron-56. (This illustration, found in one version or another in many nuclear physics texts, is to the best of our recollection a version drawn by the authors decades ago.)

This behavior has some profound effects. For example, nuclei on the low-atomic-weight part of the curve (to the left of iron) can be fused under suitable conditions to yield energy (this physical phenomenon underlies the international effort to develop light-element fusion as an energy source and is employed in thermonuclear weapons). Nuclei on the higher-atomic-weight part of the curve can undergo fission under suitable conditions, again with energy release (behavior underlying the development of various fission devices such as reactors for energy generation). In stars, fusion is the source of the energy and radiation pressure that maintain the stable configuration of such objects: four hydrogen nuclei fuse into one helium nucleus, with concomitant energy release. Fission reactions occur in the atmospheres of exploding stars. In stars that have basically used up their store of light elements, however, heavier elements will be formed as the star shrinks and heats up; according to current theory, this is the locale of the formation of the heavier elements. If the star's mass is insufficient to lead to an explosion, the heavier elements are formed basically by equilibrium reactions, in which the respective abundances of various nuclei are determined by the amount of energy they add to the mix when they are formed. If the star is of sufficient mass to explode ultimately, the equilibrium relative abundance distribution will be modified by reactions during the explosion, but evidence of the original distribution survives into the explosion products. A major feature in the evolution of stars as they run out of fusion fuel is that they evolve toward the binding energy-curve maximum in the vicinity of iron. An outstanding feature of stars going through a supernova stage is the strong presence of iron-group nuclei in the spectra of light emitted.

Gamow had a favorite way of discussing equilibrium theories: he suggested that the elements were "cooked" when they were contained at a specific temperature and density and given sufficient time to equilibrate according to the energy equivalent of the mass defect of the particular nuclei. Not surprisingly, given sufficient time, things would be sorted out in a reflection of the binding energy, with iron being the most abundant species in the mix. The most thorough exploration of an equilibrium theory was carried out in the 1940s by S. Chandrasekhar and L. R. Henrich, who concluded that the relative abundance of the isotopes of a given nuclear type (same number of protons but different numbers of neutrons), could indeed be understood in equilibrium terms, but that an overall explanation of abundances required a nonequilibrium origin. A single temperature and density specification did not suffice to explain all the elements on the heavy side of the iron peak, leading to what Gamow called the "heavy element catastrophe" (see figure 3.4), which was one of the stimuli for the development of the Big Bang model.

Steady state models. Steady State models saw much currency in the period 1948 to 1964. As mentioned earlier, the principal protagonists were the British scientists F. Hoyle, H. Bondi, and T. Gold, all of whom were or became well known. Hoyle is probably best known to the public, particularly since in the heat of a transatlantic argument on the BBC in 1949 between Hoyle and Gamow, who advocated a dynamic evolving universe, Hoyle for the first time characterized the evolving model as the Big Bang.

The original premise of the Steady State model, which has been basically abandoned since then by most cosmologists, was that the universe should on average remain the same for all time, but that it should also exhibit expansion as required by the Hubble observations. In the simplest possible terms, this would require the introduction of new matter, somehow, into the existing universe, with subsequent agglomeration of the new matter into new structures, thereby maintaining the mean density of matter and other average properties of the universe. Modifications of the Steady State model included introducing a so-called creation field to explain the appearance of new matter, or subdividing the universe into small volumes, within each of which a process not unlike the overall Big Bang model ensued. The theory foundered on a number of difficult points such as the existence and properties of the cosmic microwave background radiation, the formation of helium in abundances in agreement with observation (in the Big Bang model, helium is formed primordially), and the observed number counts of radio source entities as a function of distance (which fit the predictions of an evolving universe rather than one in a steady state). The work of Nobel Laureate Martin Ryle of Cambridge on counts of radio sources was much acclaimed as a *coup de grâce* for the Steady State, though Hoyle argued and, we believe, may still be arguing that these source counts do not rule out the model.

One positive consequence of this model is the conclusion that the observed cosmic abundance distribution of the heavier chemical elements must be the result of nuclear processes in stars. As we have already mentioned, it seems to be an unassailable fact that the light elements (hydrogen, helium, lithium and their relevant isotopes) were fashioned in the early expanding universe. With the failure of a Steady State model to provide an explanation, in 1964 Hoyle and R. J. Tayler wrote a paper on the mysterious abundance of helium in which they concluded that helium must have been formed in an early, prestellar state of the universe. This paper was widely taken as a capitulation of the Steady State advocates. We note that there is an ongoing problem in identifying the primordial abundances of deuterium and lithium, since these are readily reduced in reactions inside stars. Heavy-element formation in stars, however, was thoroughly explored in the landmark paper already mentioned by Burbidge, Burbidge, Fowler, and Hoyle. Moreover, Hoyle and, independently, E. E. Salpeter suggested an important nuclear reaction route for the formation of carbon from three helium nuclei; when this route was observationally verified by Fowler, it provided a breakthrough for modeling stellar nucleosynthesis. Stellar nucleosynthesis is still an active field of research, with increasing attention to understanding the details of relative abundances of various heavy nuclei; at the same time, light-element nucleosynthesis in the early universe also continues to receive detailed attention.

We are indebted to Albrecht for his brief but cogent refutation of the points raised by Burbidge, Hoyle, and Narlikar in their 1999 *Physics Today* and *Nature* articles which suggested an alternative to the former Steady State model. They note that if all the energy released in the formation of helium from hydrogen became available, wherever this occurred, then the resulting radiant energy might be thermalized by interaction with interstellar and intergalactic grains to a value consonant with the observed flux in the microwave background corresponding to a temperature of 2.73 kelvin. They choose to ignore the close connection between the CMBR and the extraordinary agreement of the observed abundance of light nuclei with the calculation of abundances in a primordial setting with a single free parameter. They deal with the lack of precision in measures of the abundance of He^3 and H^2 by noting that although these nuclei are significantly modified in stellar processes, it should be anticipated that future theoretical developments will rationalize their stellar production. They also suggest that a weakness in the standard Big Bang model is that dark matter has not yet been definitively observed. They also propose that their model does require some dark matter and conclude that much of it can be identified with injected new matter, as described below.

In an otherwise homogeneous, isotropic model there does not seem to be a way of forming grains (they suggest iron grains) without producing profound effects on the propagation of the CMBR. To get around this they suggest that the universe is quasi–steady state, going though a series of expansions and contractions, with thermalization by grains occurring during an appropriate phase of these cycles—and of course the present cosmos is in an expansion phase. Having thus disposed of the CMBR origin question, they go on to suppose a new mechanism for the introduction into the cosmos of the newly created matter necessary to maintain a steady density of matter, as well as to provide a source of some of the dark matter indicated by such measures as the velocity distribution of stars in galaxies, which seem to require significant amounts of dark matter. Stimulated by the arguments of Halton C. Arp, which are widely disputed by cosmologists, that his observation of galaxies in close proximity in the sky but with very different redshifts disputes the standard interpretation of redshifts, they propose that objects are ejected from a galaxy at high speed, that these objects contain newly created matter and have disparate redshifts because of the velocity of their ejection. We have a few more remarks about Arp's view a little further on.

Finally, they reject the idea that gravitational effects produce the great energy outpourings of radio sources, active galactic nuclei, and quasars but suggest instead a good case can be made that the creation process inherent in their quasi–Steady State model could provide the needed energy.

Clearly, future observations and theoretical developments will have an impact on the Burbidge-Hoyle-Narlikar model as well as on the standard Big Bang model.

Matter-antimatter models.   Hannes Alfvén and Oskar Klein were both outstanding scientists. The former was noted for his studies of the properties of conducting gases in the presence of magnetic fields, particularly in the astronomical context; the latter was well known in a number of areas of theoretical physics. Their cosmological thesis, pursued mainly in the 1950s, questioned the Big Bang model because its proponents, they said, including the present authors, had not adequately considered the problem of matter and antimatter. (In self-defense we note that in most of our early writings on the Big Bang the matter-antimatter problem was at least mentioned, usually with the caveat that a more nearly complete understanding of the physics of the very early universe would contain the necessary rationale. This has proved to be the case.)

What is the problem? It is an observational and theoretical fact that all fundamental or elementary particles have associated antiparticles. The proton has the antiproton, the electron has the antielectron (the positron), the neutron has the antineutron, and so on. Such pairs have masses in common but have opposite charge (if charged). And they have the exciting property that if a particle encounters an antiparticle, annihilation ensues, with the energy going off as radiation. It seems reasonable to suppose that particles and antiparticles could have been produced in equal numbers in the early universe. If so, perhaps there was some kind of separation, leading one to anticipate the presence in the universe of equal numbers of structures composed entirely of one or the other kind of matter. The matter-antimatter school suggested that when such matter and antimatter entities met, there would be a region of interaction where annihilation went on, and that radiation from such a region might be observable. They likened this region to the vapor separating a hot plate from an evaporating liquid next to the plate. Unfortunately, there does not appear to be any easy way for astronomers to decide whether a given galaxy is made entirely of matter or of antimatter; mass, luminosity, and spectral appearance would in theory be the same. Recent observations of incoming cosmic rays, however, have failed to find any abundance of the antiparticle equivalent of helium, which one might expect if there were significant amounts of antimatter out there.

For lack of corroborating cosmological evidence, this matter-antimatter approach has not gotten very far. Moreover, it now appears that one can understand the current preponderance of matter over antimatter as the result of symmetry-breaking in the early universe, wherein reactions among elementary particles producing matter do not proceed at the same rate as the inverse reactions, in which antimatter particles react to produce conventional particles. One other suggested possibility is that all the cosmic structures contain equal amounts of matter and antimatter, with annihilation being the luminosity-producing engine. One can argue that this cannot be the case. In fact, an abundance of as little as one part in 10 million of antimatter with respect to matter could account for all the radiation output of a galaxy, and observationally this is not tenable. Still, there are some poorly understood events in the cosmos such as gamma ray bursters; one recently observed event displayed an extraordinary amount of energy in such a burst, and no physical event has been found to explain its origin. Could this be some kind of matter-antimatter annihilation?

Plasma models. Some scientists have suggested that since so much of the universe exists in a plasma state (i.e., as ionized gases that can conduct current in the presence of a magnetic field), associated electromagnetic forces (forces between electrical currents and magnetic fields) should dominate the behavior of the cosmic structures. They go on to argue that the conventional Big Bang model must therefore be inapplicable, since it does not use this additional body force in calculating the large-scale behavior of the model. Again there is no corroborating evidence for this approach; despite a rash of technical papers dealing with the plasma cosmological model, it has now pretty much died away.

Questioning of the redshift. The next question has to do with a nonmodel rather than an alternative model of the cosmos. One of the most basic observable features of the cosmos is the redshift (an apparent Doppler effect resulting from the velocity of the light source), which is taken as a measure of the recessional velocity of the emitting object with respect to an observer. This effect shows up as the relocation of the wavelength of a spectral line; that is, radiation characteristic of a given chemical element is displaced from its position as measured in the laboratory to a longer wavelength, the displacement being proportional to distance. (We would remind the reader, at the risk of being redundant, that there is not a real Doppler shift, but rather an expansion of the space in which the galaxies are embedded.)

Some scientists, notably Halton Arp, argue that this interpretation is wrong, that wavelength shifts depend on some intrinsic property of the emitting structure and have nothing to do with a cosmic expansion. Arp came to this conclusion by noting pairs of galaxies, or a galaxy and a quasar, very near to one another in their positions as projected on the celestial sphere but exhibiting discordant redshifts. Originally he suggested that he could see a bridge of matter between such pairs, but this was later shown to be a photographic artifact deriving from forced successive development of images to bring up background features above background noise. We do not believe that study by modern, highly sensitive charge-coupled-device cameras would show any significant gradient of redshifts along these presumed bridges between galaxies, although we are not yet aware of any observational effort along these lines. Moreover, others, principally John Bahcall of the Princeton Institute of Advanced Study, have shown that the number of occurrences of these pairwise correlations of position is statistically insignificant. So the challenge to the Hubble law is dormant, except in the recent suggestions of Burbidge, Hoyle, and Narlikar.

The foregoing have been the major challenges in recent years to the conventional Big Bang.

The Singularity Revisited

Going back to study the behavior of the cosmos at a very early time in the standard Big Bang, one finds that the closer one gets to zero time, the closer the temperature and density get to infinite values, leading to a final stage called a *singularity*. This is a concept generally abhorred in physics, yet some years ago Stephen Hawking and Roger Penrose developed the interesting result that any cosmological model depending entirely on the general theory of relativity would exhibit a singularity. In addition to the one predicted by the Big Bang model, there must be one at the center of a black hole, a collapsed stellar object whose mass concentration is so high that even light can no longer escape. If it seems unreasonable to expect an observable singularity in any real physical system, some change in the relevant physics of the system is needed to avoid the singularity.

Einstein remarked that the general theory of relativity should not be expected to apply to a singularity. In fact, some of the current ideas about inflation at very early times avoid consideration of a singularity by invoking a limiting size for the radius of the curvature of space: namely, the Planck length of 10^{-33} centimeters. An example of where physics overrides the possibility of a singularity might be the white dwarf star. Such a star has exhausted its nuclear fuel, and so there is no longer the pressure of radiant energy to prevent the star from collapsing under gravity. Collapse it does, to an extent where the nuclear constituents basically merge, establishing a new kind of pressure that prevents further collapse. This is called *degeneracy* pressure. It results from the fact that electrons freed from the compressed nuclei can occupy only certain energy states—a quantum mechanical effect—and in essence become incompressible; thus a singularity is avoided. What the introduction of appropriate physics will do for the more general cosmological singularity is not yet known, but some of the current approaches to modeling the very early universe with grand unified theories and inflation will very likely lead to a suppression of the singularity.

The Cosmological Entropy per Baryon

We alluded earlier to the fact that in theoretical studies of the production of light elements in the early expanding universe, one quantity must be assumed because it cannot yet be predicted from first principles: the density of matter at the time of element production. It is convenient to express this density as the ratio of the number of particles present to the number of photons (uniquely determined by the temperature, which is in turn uniquely determined by the cosmological model in which radiation dominates at early time). In successful representations of the light-element abundances this ratio is found to be of the order of 10^{-9} to 10^{-10}.

This size is surprising to many, since the photon density is related to the entropy of the universe and is indeed quite large.

There is no magic in the value of this ratio. The early universe is dominated by radiation, behaving as though it is blackbody radiation, for which the Planck energy distribution can be evaluated to yield the number density of photons. The result is that $n_{ph} = 20.3 \, T^3$, where 20.3 is a dimensional number that comes out of theory. On the other hand, the density of matter drops as the volume increases, and the volume varies as the cube of the distance, l, between any two arbitrarily chosen points in the cosmos. In a radiation-dominated universe, however, the temperature T varies as $1/l$. Therefore, the ratio $n_{\text{ph}}/n_{\text{particles}}$ depends on T^3/T^3 and is completely independent of the expansion. This ratio is thus a constant containing the density of matter as a factor, a factor that is chosen to give the proper abundances of elements in the short duration of primordial nucleosynthesis.

This ratio has some other interpretations. Consider its inverse. Numerically, the number density of photons divided by the number density of particles is about 1/7 of the ratio of the heat capacities of radiation to matter. It is also 1/8 of the ratio of pressures of radiation to matter, and 2/7 of the dimensionless entropy per baryon. These quantities are relevant in the attempt to interpret the high value of entropy in the early universe. Some scientists have related this high value to entropy generated very early on by the processes that lead to particle generation. We mentioned earlier that Alpher and Marx considered the generation of entropy in the early universe if the universe were closed (i.e., went through successive states of expansion and contraction—an oscillating model, if you will) and found that successive collapses and expansion would increase the entropy of the universe each time. The entropy in this view arose from a small exchange of energy between matter and radiation.

There are both philosophical and physical quandaries associated with questions of entropy in the universe. The textbook definition of entropy can be expressed in terms of classical thermodynamics or equivalently in terms of statistical mechanics. Since discussing this definition in detail would surely lose much of the audience, suffice it to say that entropy is a measure of the disorder in a system. It has a maximum value for any physical system that is in equilibrium, a system whose components are on average uniformly distributed. An interesting sidelight is that when any system is in equilibrium, it is in principle not possible to say anything about its previous history. In the cosmological context, if the early universe (during the radiation-controlled early phases) was at any time in a state of equilibrium, then one might not be able to say anything in detail as to how the cosmos arrived at that state. This conclusion has always given us a bit of heartburn in trying to take seriously discussions of inflation in the early universe, despite the fact that inflationary paradigms do indeed address some of the conundrums of the established standard Big Bang model which does appear to have gone through an equilibrium state prior to the formation of light elements.

To continue, note that as any physical system evolves, it is the case that the entropy is unchanged (in a system in equilibrium) or increases (in systems not in equilibrium). Another interpretation is that entropy is a measure of the number of possible physical states available and accessible to a system. If, in the early universe, we have processes that produce particles, this leads to an increase of entropy. It is a powerful concept in physics. (Recall our reference to Ilya Prigogine's proposal of a close connection between particle production and the production of entropy.)

Clearly, there is still much to be learned, but as of this writing, there appears to be no significant challenge to the Big Bang model.

8

The
Future of the
Universe

Prediction is very hard, particularly
of the future.

—Neils Bohr

Hard as prediction may be, and certainly beyond the bounds of verification, there are some things we can say about the future of the universe.
First, anything we say is predicated on the assumption that the laws of
nature will remain the same in the future as they are now, that the behavior of matter and energy is the same at great distances from us as it
is locally and therefore was the same in the past, and that there will be
no new laws cropping up. One can say these things with some confidence,
since it has been demonstrated by studies of the past behavior of the
universe that, where checking is possible, these laws were the same in
the past, virtually back to the Big Bang event itself. If one accepts the
premises of the standard Big Bang model, and in particular that the universe is homogeneous and isotropic when viewed on a sufficiently large
scale, and moreover, if one ignores structure so that the universe is considered to be filled uniformly with matter and radiation, then there are
consequences for the future which merit discussion.

The general relativistic model that is the basis of modern cosmological modeling in the standard Big Bang provides three possibilities for the
future gross behavior of the universe, behavior that we have discussed

briefly earlier. The particular behavior depends on the total density of matter and radiation compared with a so-called critical density. This ratio, Ω, can in mathematical terms have any value but in practical terms appears to lie between a few tenths and two. If Ω lies outside this useful range, one finds, on the one hand, that for large Ω the speed of nuclear reactions in the prestellar universe is too high to account for the requisite number of light nuclear species; moreover, any reasonable calculation of the age of the universe yields ages that are short compared with the imputed ages of the various structures in the universe. On the other hand, if the value of Ω is too low, then there has not been enough time in any reasonable age of the universe to produce the observed structures.

The critical density is that density which serves to divide the possible cosmological models into either one where $\Omega > 1$, the so-called closed model (also called a spherical or elliptical model), or one where $\Omega < 1$, the so-called open model (also called a hyperbolic model, with an underlying saddle shape). The possibility that $\Omega = 1$ describes a flat universe: one which does not exhibit any curvature of space arising from the distribution of matter and radiation and which therefore behaves like a Newtonian universe with Euclidean geometry and no dependence on the amount of matter and radiation. With $\Omega < 1$ or $\Omega > 1$, the geometries are no longer Euclidean.

The flat universe case, $\Omega = 1$, is favored by many cosmologists for several reasons. One is that to some it appears to be a particularly elegant choice, even though it separates possibilities in a very narrow range between $\Omega > 1$ and $\Omega < 1$. It may be argued that since observation places the value of Ω close to unity, why would not the $\Omega = 1$ description be the logical choice? We have found it interesting that in our own work on the standard Big Bang model and in some of the standard textbooks on cosmology, the mathematical approximations for Ω at very early times give a value exceedingly close to unity for some hundreds of thousands of years into the expansion (see the appendix). For whatever reasons one may want a value of $\Omega = 1$ in the early universe, there it is, with or without inflation. The proximity of Ω to unity in the standard Big Bang model covers the time during which inflation would have gone on and for a considerable time thereafter; thereafter Ω begins to deviate from unity and can easily encompass a value different from unity later on should observations require it. What is observed gives a value of Ω at present, based on the amount of conventional matter (baryons), which is of the order of a few tenths. The most important reason for favoring $\Omega = 1$ is that it appears to be demanded by inflationary paradigms for the behavior of the early universe. Moreover, there are other possible contributions to an overall cosmic value of Ω. We feel that a present value of Ω different from unity is or could be acceptable. We say more about these observations below, particularly since they are relevant to discussions of the future.

It is customary to deal with the three possible solutions by examining the time evolution of the distance between any two points, however such points are selected. Changes with time in this distance, which relativists call proper distance, are clearly related to a scale factor for the expansion, which scale factor describes how this arbitrarily selected distance behaves with time. The three possibilities are shown in figure 6.1, a schematic representation of the behavior of the scale factor. Let us reiterate what we said earlier about the three cases. If the universe is open, with $\Omega < 1$, then the scale factor ultimately becomes asymptotic to a constant expansion rate, and the cosmos will continue to expand forever. If the universe is flat, then the scale factor asymptotically approaches a constant value, with the expansion rate equal to zero. Finally, if the universe is closed, with $\Omega > 1$, the universe will expand to some maximum value of the proper distance where the expansion rate will go to zero, and then the expansion will change sign with the universe going into what has been called a Big Crunch, returning in time to a hot, highly dense initial state (i.e., to the early Big Bang), with possible subsequent oscillatory behavior.

The geometries of the three cases differ in interesting ways. In an $\Omega = 1$ model, which displays Euclidean geometry, parallel lines remain parallel forever, the sum of the angles inside a triangle is 180°, and the circumference of a circle is 2π times the radius. In an $\Omega > 1$ model, lines that start out parallel ultimately cross, the angles inside a triangle add up to more than 180°, and the circumference of a circle is more than 2π times the radius. And, as one might guess by now, in an $\Omega < 1$ model parallel lines do meet ultimately, the sum of a triangle's angles is less than 180°, and the circumference of a circle is less than 2π times the radius. To expand on the idea of parallel lines meeting, think of a pair of lines at the earth's equator which are parallel. Now monitor the separation of this pair of lines as you move toward one of the earth's poles. There they meet and cross. These several geometries are illustrated in figure 8.1, where we call attention to the three cases: being flat, spherical (also called closed, a hypersphere or elliptic model), and like a saddle surface (also called open or hyperbolic).

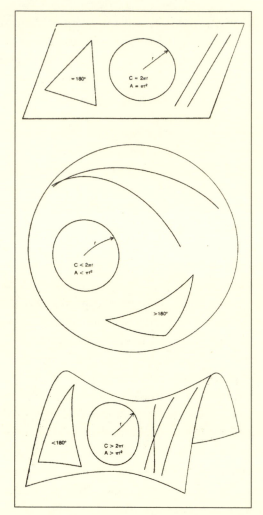

Figure 8.1. Representation of the three basic geometries in the standard Big Bang model. The top drawing shows a flat or Newtonian universe, which is defined by the density parameter Ω being unity. Shown on the figure are the circumference of a circle, C, the area within the circle, A, the sum of the number of degrees in the three vertices of a triangle, and the behavior of parallel straight lines, which maintain constant separation. The middle drawing shows the spherical (sometimes called elliptical or closed) geometry associated with the density parameter Ω exceeding unity. The inserted descriptions are, as before, C, A, total of degrees in a triangle, and parallel lines that do meet somewhere. The bottom drawing shows a saddle surface (sometimes called negative, hyperbolic, or open) for a density parameter Ω less than unity. The descriptors are as in the other drawings. (To the best of our recollection, this figure was produced by Alpher decades ago, although there are similar diagrams in cosmology texts.)

So there are three possible solutions of the equations of the Big Bang model. Now briefly consider the future of the contents of the universe, the various structures it contains: gas, dust, stars, stellar clusters, galaxies, clusters of galaxies, and of course—a feature common to the three models—the cosmic microwave background radiation. This last has a current value of 2.73 kelvin. So long as there is no significant net energy transfer into the background radiation, in either the open or the flat model, the radiation should on the whole continue to expand and cool. New structures may form, and older stars may snuff out, but the universe should continue to cool as it expands. Some lighter stars will die, go into a white dwarf stage, and ultimately snuff out. Or, if the white dwarf is part of a binary star system in which the other member remains more or less normal, there will be material transfer from the "healthy" star to the white dwarf, which will ultimately become a supernova, and thus hasten its final fate as a cooling ember. Heavier stars will go into a collapse phase and become either supernovae or black holes. The supernova will become a neutron star and radiate its residual energy; the black hole will decay and ultimately vanish by emitting particulate radiation (called Hawking radiation, since Stephen Hawking originally predicted this as yet unobserved happening). Galaxies will decay by the loss of stars, dust, and gas held by gravitational forces, and clusters of galaxies will lose the structural coherence they possess as a result of gravitational binding. This makes sense when one realizes that all these entities have velocities that buck the force of gravity and, from time to time, may get enough energy transferred in encounters with other entities to allow them to exceed an escape velocity. In the Milky Way, for example, it is clear that almost all stars have a component of velocity perpendicular to the plane of the Milky Way disk, and in the fullness of time these objects will evaporate from the Milky Way.

The closed universe may or may not go through the foregoing stages, depending on how quickly it reaches the time when the expansion stops, and the universe begins its return trip toward its hot, dense original state. In all of this, one should keep in mind that while stars and galaxies are evolving and dying, the dying supernovae will distribute newly synthesized heavier elements in space, from which new stars and perhaps galaxies can form. But ultimately, the availability of nuclear fuel will cease, and that will be that. Only the closed universe has another chance to do it right or at least differently the next time.

It is possible to apply the known laws of physics to make some quantitative estimates of the time scale of the various events that may occur. Excellent review articles for the more technically trained reader and several books (notably, Islam and Morris) on the subject of the future of the universe are listed in our recommended readings. In their 1997 review paper, listed in the appended reading list, Adams and Laughlin systematically examine the future of planets, various kinds of stars, stellar populations, galaxies and galactic clusters, the cosmic background radiation, and the cosmos generally. (This paper has been expanded into a book, which we have not yet seen.) It is interesting that to a considerable extent, details of the future depend on whether the proton is a radioactive nucleus. Measurements thus far indicate that the lifetime for decay of the proton is in excess of 10^{32} years, which is certainly a good long time, suggesting that for the very long-term future the proton is rock solid. Adams and Laughlin, taking 10^{37} years as a canonical figure for proton decay, provide a timetable summarizing their calculations and final conclusions about the future.

We choose instead to mention some of the problems listed by Adams and Laughlin which affect profoundly any view of the future. First and foremost is the question of the radioactive decay of the proton. It is important not only for considerations of the future but for efforts to understand the unification of forces; theories that say something about the decay of the proton can be sorted out depending on the decay rate. A second question of importance is whether the vacuum energy density of the universe is nonzero. If so, then one may expect an acceleration of the cosmic expansion at early times, or there may be a future period of inflation once more. If the vacuum energy density is really zero, then one cannot rule out a closed universe and a future Big Crunch.

A third profound question is the nature of the dark matter. What is it, and is it radioactive, with a finite lifetime? A fourth is the competition between the evaporation of stars from galaxies into intergalactic space and their absorption by the black holes that appear to inhabit the center of galaxies. This is of particular interest because black holes have finite lifetimes for mass loss by Hawking radiation, which depends on black hole mass. Finally, is something weird to be expected when the universe has expanded and cooled to very low temperatures, and when the wavelengths of blackbody photons become comparable to the dimensions of the visible universe?

As residents of the third planet from the Sun, a star that will surely go into an evolutionary phase when its primordial stock of hydrogen for fusion energy production becomes diminished, we should perhaps be concerned about when the Earth will be engulfed by the expanding envelope of the red giant—although, alternatively, the Sun may ultimately lose a good fraction of its mass through stellar winds and never engulf Earth in its atmosphere. Suffice it to say that these options lie billions of years in the future, and we should not be worried in the short term.

9

The
Anthropic Principle

We have said a good deal about alternatives to the Big Bang, particularly in chapters 4 and 7, but feel impelled to say a bit more, particularly about the anthropic cosmological principle. We do not regard it as an alternative to the Big Bang model, but there has been so much written about it, at both technical and popular levels, that we must deal with it at least briefly.

Some writers deal with the basic idea that the universe must contain equal amounts of matter and antimatter; this notion has been at least qualitatively concluded to be an unneeded complication, since an inequality of relative abundances of matter and antimatter is a concomitant of theoretical studies of elementary particle reactions in the very early universe. Some deal with the idea that since much of the material we see in the universe is in a plasma state (ionized material that can transport electrical currents), electromagnetic phenomena must determine the structure of objects we see. Forces that arise from electromagnetic phenomena are undoubtedly important, but at present the basic structures are explainable in terms of gravitational forces. Other writers introduce interesting but inappropriate explanations of the properties of elementary particles in terms of vortices, multidimensional structures, or other imagined structures. Still others simply reject the basic premises of the Big Bang model, primarily on philosophical grounds, arguing that there must be some other approach to understanding the nature of the universe, given the model's enigmas. Finally, many authors have insisted on literal interpretation of the cosmologies contained in those writings regarded as divine revelation in various religions.

So we now presume to say something about an aspect of cosmology known as the anthropic cosmological principle, which some may consider pseudoscience but which in fact proceeds on a completely different level of analysis. By and large, those who espouse the anthropic cosmological principle are otherwise productive scientists involved in cosmological research, who do not deserve to be ignored or demeaned by anyone. The principle is a philosophical approach to understanding the cosmos which accepts the basic premises of the Big Bang model, an approach to which Brandon Carter, Robert Dicke, John Wheeler, John Barrow, and Frank Tipler, among others, have made significant and serious contributions. Barrow and Tipler's 1986 book, *The Anthropic Cosmological Principle*, which we recommended in chapter 4, aside from espousing their ideas, is an excellent overview of the state of the Big Bang model at the time they wrote it (and it provides references to other important works). Although we disagree with many of the authors' interpretations, it is nevertheless an interesting and thought-provoking treatise, as well as a good historical review, and parts of it should be accessible to readers without a sophisticated background in cosmology.

In order to convey the rationale for the introduction of the anthropic principle, we propose to proceed from some of the ideas put forth by reviewers of Barrow and Tipler's book and as well from a specific piece of work carried out by Gamow and Alpher.

In its contemporary form, the principle appears to have originated in suggestions made by G. J. Whitrow, Brandon Carter, and Robert Dicke in the late 1950s. Whitrow had the interesting suggestion that the universe had to have three spatial dimensions in order for life to exist. As early as the work of Copernicus, it had been almost axiomatic in cosmological modeling that the observer does not occupy a privileged position in the universe; that is, any observer, wherever and whenever located in the universe, would see the cosmos to be the same everywhere and in all directions, as would a terrestrial observer. But Carter introduced the idea that our location in the universe is in fact necessarily privileged to the extent of being compatible with our existence as observers. Putting it differently, Dicke said that the universe must be so constructed as to allow humankind to have evolved; the basic features of the cosmos must include enough elapsed time and a suitable physical and chemical environment, including the presence of those chemical elements—particularly carbon, hydrogen, oxygen, and nitrogen—necessary for our form of life. To put it somewhat differently again, observers must exist (with all this connotes in biology and the theory of evolution) in order that the fundamental constants of nature be measurable. These remarks constitute the so-called weak anthropic principle, which even Barrow and Tipler consider to be somewhat tautological.

Carter went a bit further in arguing that life of any sort would be impossible unless the basic laws of nature were exactly what they are. There is much one can say, or would like to be able to say, about the constants of nature: about why the masses of elementary particles have the values they do, for example, and about the apparent relationships between the dimensionless ratios one can develop using the various constants of nature, as with Paul Dirac's "law of large numbers" (about which we will shortly say more). One can be misled by the novelty of these dimensionless ratios; for example, one set of ratios proposed by Dirac involves the age of the universe, which then requires that something else must vary with the age of the universe in order to keep such ratios constant. It is possible to show that one can with seemingly equal validity construct ratios that do not depend on time-varying constants of nature.

One can also imagine that there are many different universes, each characterized by a different set of natural constants which may or may not be compatible with some form of life, and that we live in one such option. Scientists espousing the anthropic cosmological principle are also intrigued by the very considerable size of the entropy per baryon, discussed earlier. This ratio, which has a numerical value required by the theory of primordial nucleosynthesis in order that calculations agree with observed cosmic relative abundances, is thought by some to demand a unique cosmological model.

Wheeler went on to propose what he called a participatory anthropic principle: namely, that no universe can exist unless it contains conscious observers. This view may have a deep connection with quantum mechanics in that it makes the observer an essential part of the larger system—that is, the cosmos. The universe does not exist unless and until it (where "it" includes observers) can study itself through its observers.

These views lead to the final anthropic principle, which is simply that since life exists on earth, then—if there is no life elsewhere in the universe—terrestrial life cannot be destroyed, for the destruction of the observers as part of the system would also wipe out the universe, which requires observers in order to be a real entity. As this is written, we note that more than forty stars have been shown to have a planetary body in orbit, and in the recent past a star called Upsilon Andromedae has been shown to have at least three planets, all as large as or larger than our planet Jupiter. It is difficult to imagine that there are no smaller planets in this system, whose gravitational effects on the star would be small and below the limit of present observability.

Within the main thrust of ideas about the anthropic cosmological principle, there is much dependence on the constants of nature for hints as to the validity of the premise. The variety of such quantities and of the ratios one can form from them are admittedly intriguing, although some have already been diminished in import by recent observational studies. For example, it was suggested some years ago by Vittorio Canuto and colleagues that there might be two time scales governing cosmic phenomena. One scale would have been determined by timekeeping based on the regularity of the orbits of planetary bodies in the solar system—that is, would be conditioned by the laws of Newton and Kepler (and general relativity in the case of Mercury, to explain the precession of its perihelion). The other time scale would arise from periodicity in atomic and molecular phenomena—that is, would be conditioned by quantum mechanics. This interesting suggestion has been found wanting by observations made of the travel time of signals from a satellite near Mars. We have already noted that the Dirac large-number hypothesis seems to fail because it introduces a time scale for cosmic phenomena based on the age of the universe, which is certainly model-dependent and can be obviated by using some other time as characteristic of the model: for example, as suggested by Alpher and Gamow, the time in the expansion when the densities of matter and radiation were equal. This provides a characteristic time that does not vary with the epoch. For those interested in pursuing the apparent coincidences and relations among the constants, there is no better place to start than with the book by Barrow and Tipler or, at a more elementary level, *The Symbiotic Universe* by George Greenstein (New York: Morrow, 1988. See esp. pp. 58, 78).

Details aside, it seems to us that the criticisms of the anthropic principle given by Heinz Pagels in his book *Perfect Symmetry* (New York, Bantam Books, 1985) are most cogent for scientists and for those who adhere to the scientific process: the anthropic principle does not predict anything and is rather "a far-fetched explanation for those features of the universe which physicists cannot yet explain" (p. 377). We see the anthropic cosmological principle as a blind alley off the road to understanding the nature of the cosmos, an alley to be explored yet far from being taken seriously by most cosmologists. It must be viewed with considerable care because it lacks those characteristics that we usually associate with scientific theories. To quote Harrison's book *Cosmology* (see our recommended reading list), "We shall occasionally refer to the anthropic principle, and the reader may, if it is preferred, substitute the alternative theistic principle" (p. 113).

Let us now return to the particular aspect of the arguments concerning the fundamental constants which was the subject of the final research work done by Alpher and by Gamow just before he died. Gamow's interest was piqued by, among other things, a general concern about connecting physics in the small (particle, atomic, and molecular physics) and physics in the large (cosmic structure and evolution). Having established for the first time a relation between the synthesis of the chemical elements and the properties of the very early universe as a milieu for nucleosynthesis, we were intrigued by the idea developed by P.A.M. Dirac which led him to suggest his law of large numbers: to wit, "any two of the very large dimensionless ratios occurring in nature are connected by a simple mathematical relation in which the coefficients are of the order of magnitude unity. Moreover, if one or more of these relations vary with the epoch (say, the age of the universe), then so must they all." What are these ratios, typically? Dirac proposed the following examples.

Forces	$\dfrac{\text{Electrostatic (Coulomb) forces between an electron and proton}}{\text{Gravitational force between these particles}}$
Length	$\dfrac{\text{Characteristic scale in the universe}}{\text{Classical radius of the electron, or the range of forces between nuclei}}$
Masses	Mass of the visible universe = Number of particles in the universe/Mass per particle
Energy	$\dfrac{\text{Gravitational potential energy of rest of universe in the field of a nucleon}}{\text{Rest of mass energy of a nucleon}}$

Each of these dimensionless ratios contains as a factor either the constant of gravitation, G, as first introduced by Newton, or the age of the universe expressed in atomic time units; and each of the ratios has a numerical value of $(10^{40})^n$, where n is 0, 1, 2, or 3. For this to be so, Dirac suggested that G must vary inversely as the epoch. This result led to a spate of work on new approaches to cosmology, including those of Pascaul Jordan and his colleagues in Germany, Robert Dicke and Carl Brans, and many others, looking at the effect of a time-varying gravitational constant.

It seemed to Gamow and Alpher that if one introduced the age of the universe as a characteristic time and insisted that the ratios be constant, then other attributes of the cosmos must also vary with the epoch. Among other approaches that also occurred to us, the main one was that there is a characteristic time in the universe which is fixed once and for all by the parameters of a given model: namely, the time of crossover or decoupling, when the behavior of the universe went from being radiation-controlled early to being matter-controlled thereafter. If now we revisit Dirac's hypothesis, make this change in characteristic time, and make one other change in the force ratio—namely, take as fundamental the force between two protons, rather than the force between a proton and electron—then the dimensionless ratios all have the value $(10^{36})^n$. It is amusing that now the ratio of specific heats of radiation and matter has a similar dependence, which was not the case with Dirac's ratio. Before he died, Dirac wrote to Gamow and Alpher that in the end we should look for observation to settle this issue, and it now appears that recent measurements have ruled out any significant variation of G with time.

Epilogue

Science is self-correcting and progressive, and its output is cumulative. Thus the work of a Newton is subsumed later in the work of an Einstein. Newton was not wrong; rather, his world view was limited by the knowledge in his time.

Science brings some problems upon itself. For example, scientists are frequently attracted by a theory because they find it beautiful, when in fact it may be wrong; our colleague George Gamow always wondered if there was something worth pursuing in Dirac's law of large numbers, which he found to be beautiful. And from time to time a paper appears in the scientific literature in which the authors have reinvented the wheel. All this reflects the fact that the doing of science is a human and creative process.

Although many questions about cosmological modeling are still unanswered, the Big Bang model is in reasonably good shape. We are certain that future theoretical and observational work will at the very least fine-tune it, but we do not anticipate that, after more than 50 years, the model will turn out to be basically inadequate. Would that we could come back after another 50 years and see how it all came out.

Appendix

Line Elements and the Energy Equation

The Friedmann-Lemaître-Robertson-Walker line element for a homogeneous, isotropic expanding universe can be written as

$$ds^2 = -\frac{e^{g(t)}}{\left(1 + \dfrac{r^2}{4R_0^2}\right)}(dr^2 + r^2d\theta^2 + r^2\sin^2\theta d\phi)^2 + c^2dt^2 \quad \text{cm}^2 \quad \text{(FL-1)}$$

where r, θ, and ϕ are spherical polar coordinates, with the origin of coordinates being any observer; $e^{g(t)}$ defines the time dependence of the expansion and is the square of a scale factor $S(t)$ for the expansion; R_0 is the unit of measure of radius of curvature; and ds is the length of an infinitesimal interval along a trajectory in the universe. Note that in ordinary static 3-space one has

$$ds^2 = dr^2 + r^2d\theta^2 + r^2\sin^2\theta d\phi^2 \qquad \text{(FL-2)}$$

or, equivalently,

$$ds^2 = dx^2 + dy^2 + dz^2$$

For radial expansion in the nonstatic case, we can transform (FL-1) into a useful form

$$ds^2 = -S^2(t)\left[\frac{dr^2}{(1+kr^2)^2} + r^2 d\theta^2 + r^2 \sin^2\theta d\phi^2\right] + c^2 dt^2 \quad \text{cm}^2 \qquad \text{(FL-3)}$$

where for R_0 real, infinite, or imaginary one selects $k = +1$, 0, and -1, respectively.

If the cosmic medium is a perfect fluid of density ρ, which is free of viscous shear, then from Einstein's field equations we can derive an energy equation which describes the time rate of change of a distance ℓ between two arbitrarily chosen coordinate points

$$\frac{d\ell}{dt} = +\sqrt{\frac{8\pi G}{3}\rho\ell^2 - \frac{c^2\ell_0^2}{R_0^2} + \frac{\Lambda\ell_0^2 c^2}{3}} \quad \text{cm sec}^{-1} \qquad \text{(FL-4)}$$

where the $+$ sign denotes expansion, c, G, ℓ_0, and R_0 are the velocity of light, the constant of gravitation, the unit in which proper distance is measured, and the unit of measure for the radius of curvature, respectively; Λ is Einstein's cosmological constant. We rewrite (FL-4) by using

$$L = \frac{\ell}{\ell_0} = \frac{1}{1+z} \qquad \text{(FL-5)}$$

where z is the redshift. We pick an ℓ_0 such that $\left(\dfrac{\ell}{\ell_0}\right)_{\text{now}} = L = 1$, consonant with zero redshift nearby. Then

$$\frac{1}{L}\frac{dL}{dt} = +\sqrt{\frac{8\pi G}{3}\rho - \frac{c^2}{L^2 R_0^2} + \frac{\Lambda c^2}{3L^2}} \quad \text{sec}^{-1} \qquad \text{(FL-6)}$$

Note that $\dfrac{1}{L}\dfrac{dL}{dt}$ is the Hubble parameter (also called the Hubble constant) and has its present value H_0 when $L = 1$. Research for a more precise value of H_0 is a very active field in observational cosmology. If the medium is a mix of noninterconverting matter and radiation, $\rho = \rho_m + \rho_r$, then conservation of matter requires

$$\rho_m L^3 = \text{constant} \qquad \text{(FL-7)}$$

while invoking adiabatic expansion of radiation (see the discussion of cosmic microwave background radiation in chapter 5) yields

$$\rho_m L^4 = \text{constant} \qquad \text{(FL-8)}$$

and the useful expression

$$\rho_r \rho_m^{-4/3} = \text{constant} \qquad \text{(FL-9)}$$

follows, one which applies throughout the expansion, given homogeneity and isotropy.

This equation makes it a simple task to calculate the background temperature in the universe (see the section in the appendix on cosmic microwave background radiation).

If ρ_r dominates at early times

$$\rho_r \gg \rho_m$$

and $\dfrac{c^2 \Lambda}{3L^2}$ and $\dfrac{c^2 \ell_0^2}{R_0^2}$ are small, then

$$\rho_t = 4.48 \times 10^5 t^{-2} \text{ gm cm}^{-3} \qquad \text{(FL-10)}$$

and using the Stefan-Boltzmann law, $\rho_r = aT^4/c^2$ (see the section in the appendix on cosmic microwave background radiation)

$$T = 1.52 \times 10^{10} t^{-1/2} \text{ kelvin} \qquad \text{(FL-11)}$$

where t is in seconds and the numerical coefficient contains only universal constants. Using the Stefan-Boltzmann relation between ρ_r and T we can write

$$\rho_{r0} = 0.84 \times 10^{-35} T_0^4 \text{ gm cm}^{-3} \qquad \text{(FL-12)}$$

We return now to (FL-5) and consider what happens when $R_0 = \infty$. This is the case of a flat universe, Newtonian in behavior. We consider in particular (FL-5) when $t = t_0$ (i.e., now), when $L = 1$. Then

$$\left(\frac{1}{L} \frac{dL}{dt} \right)^2_{\text{now}} = H_0^2 = \frac{8\pi G}{3} \rho \qquad \text{(FL-13)}$$

and

$$\rho = \rho_{\text{critical}} = \rho_{cr} = \frac{3H_0^2}{8\pi G} \tag{FL-14}$$

It is convenient to write a density ratio as

$$\Omega(t) = \frac{\rho(t)}{\rho_{cr}} = \frac{8\pi G}{3H_0^2}\rho(t) \tag{FL-15}$$

We refer back to (FL-5) and note that we can deal with the radius of curvature R_0 unspecified and solve for R_0

$$R_0 = \frac{c}{\sqrt{\Omega - 1}}\frac{c}{H_0} \text{ cm} \tag{FL-16}$$

Some useful relationships follow from all of the above. Let T_0 be the cosmic microwave background temperature now, and let ρ_{m0} be the present smeared-out density of matter in the universe. Then the entropy per baryon (basically the entropy of radiation per baryon, with n the concentration or number density of baryons) is

$$\frac{S_r}{nk} = 1.22 \times 10^{-22}\frac{T_0^3}{\rho_{m0}} = \text{ constant during expansion} \tag{FL-17}$$

and k is the Boltzmann constant. For an ideal gas $\rho_m = nkT$. The ratio of radiation to matter pressure is

$$\frac{p_r}{p_m} = 3.6 \times 10^{-22}\frac{T_0^3}{\rho_{m0}} \tag{FL-18}$$

The number density of black body photons is

$$n_\gamma = 20.3\, T_0^3 \text{ cm}^{-3} \tag{FL-19}$$

And the ratio of photon number density to baryon number density is

$$\frac{n_\gamma}{n} = 3.40 \times 10^{-23}\frac{T_0^3}{\rho_{m0}} \tag{FL-20}$$

By way of example, if $\rho_{m0} = 10^{-31}$ gm/cm^3, and $T_0 = 2.725$ kelvin

$$\frac{S_r}{nk} = 2.5 \times 10^{10}$$

$$\frac{p_r}{p_m} = 6.3 \times 10^{10}$$

$$n_\gamma = 420 \, \text{cm}^{-3} \qquad \text{(FL-21)}$$

$$\frac{n_\gamma}{n} = 6.9 \times 10^9$$

Returning now to the earlier equations in this section, recall that if $\Omega_0 > 1$, then R_0 is real. If $\Omega_0 = 1$, then $R_0 = \infty$, and if $\Omega_0 < 1$, R_0 is imaginary. These three cases were illustrated in the text, where we plotted the time behavior of the scale factor $R(t) = L/L_0$ versus time since the Big Bang. It is useful to write ρ_{cr} as follows. Let

$$H_0 = 100h \text{ km sec}^{-1}\text{Mpc}$$

where

$$1 \text{ Mpc} = 3.09 \times 10^{24} \text{ cm}$$

Then one can write

$$\rho_{cr} = 1.88 \times 10^{-29}h^2$$

The Einstein Static Model

To obtain the static model of Einstein, it is convenient to write a line element that is spherically symmetric about any origin of coordinates (i.e., any observer), as

$$ds^2 = -e^\lambda dr^2 - r^2 d\theta^2 - r^2 \sin^2 \theta d\phi^2 + e^\nu c^2 dt^2 \text{ cm}^2 \qquad \text{(E-1)}$$

where λ and ν are functions of r. If we write Einstein's field equations for pressure and density, utilizing this line element, then we get

$$\frac{dp}{dr} = -\frac{\rho_c^2 + p}{2}\frac{dv}{dr} \qquad \text{(E-2)}$$

But ρ must be the same everywhere, so that $\dfrac{dp}{dr} \equiv 0$. Thus

$$\left(p + \rho c^2\right)\frac{dv}{dr} = 0 \qquad \text{(E-3)}$$

If $dv/dr = 0$, then $v = $ constant and (E-1) reduces to the line element in special relativity. More generally, we can rewrite (E-1) as

$$ds^2 = -\frac{dr^2}{1 - r^2/R^2} - r^2 d\theta^2 - r^2 \sin^2\theta d\phi^2 + c^2 dt^2 \qquad \text{(E-4)}$$

where

$$\frac{1}{R^2} = \Lambda - \frac{8\pi G}{3}\frac{p}{c^2} \qquad \text{(E-5)}$$

If ρ is small, then

$$\frac{1}{R^2} = \Lambda \qquad \text{(E-6)}$$

which is flat space-time.

If the model is to contain any matter at all, then $\Lambda > 0$ is needed. The model exhibits no redshift, real or otherwise.

The De Sitter Model

Return to eq. (E-2) and note that one can also have

$$p + \rho c^2 = 0 \qquad \text{(deS-1)}$$

rather than $dv/dr = 0$. One can show in this case that Λ and v in (E-1) have values such that

$$\frac{1}{R^2} = \frac{\Lambda + \frac{8\pi G}{3}\rho}{3} \qquad \text{(deS-2)}$$

However, since $\rho \geq 0$, however small it may be, then we must have $p \geq 0$. This then leads to the requirement

$$p = \rho c^2 = 0 \qquad \text{(deS-3)}$$

(i.e., an empty universe) with, again

$$\frac{1}{R^2} = \frac{\Lambda}{3} \qquad \text{(deS-4)}$$

It is interesting that although the De Sitter model is empty, there is nevertheless a calculable redshift for test photons, reflecting differences in proper (local) time at positions of emission and detection.

Cosmic Microwave Background Radiation

The cosmic background radiation is now known with great precision to be blackbody in nature, exhibiting a temperature of 2.725 kelvin, with its peak intensity in the microwave portion of the electromagnetic spectrum (COBE spectrum in figure 5.2). Superimposed on the remarkable Planck spectrum are tiny but real fluctuations of the order of a part in 100,000, fluctuations of as yet unknown origin which formed in the early universe; they are undoubtedly key to subsequent formation of structure, and are currently the subject of active study. Consider some key relationships. If matter is conserved in any chosen volume element in the universe, then the density of matter is given by

$$\rho_m(t)L^3 = \rho_{m0} \tag{C-1}$$

where L is the scale factor of the expansion, in units such that $L = 1$ now, where $\rho_m(t) = \rho_{m0}$ and with smaller values at earlier times, and the black-body radiation has a mass-density equivalent according to the Stefan-Boltzmann law, given by

$$\rho_r = \frac{4\sigma}{c^3}T_r^4 \text{ gm cm}^3 \tag{C-2}$$

Here σ is a numerical constant called the radiation density constant, c is the velocity of light, and the temperature of the radiation is T in kelvin (°K). In a blackbody enclosure the pressure of radiation is

$$p_r = \frac{1}{3}\rho_r c^2 \text{ dynes/cm}^2 \tag{C-3}$$

If the enclosure has a volume V, then the total radiant energy in the enclosure is

$$E_r = \frac{4\sigma}{c}VT^4 \text{ ergs} \tag{C-4}$$

If we combine (C-1) and (C-2) appropriately,

$$\rho_r \rho_m^{-\frac{4}{3}} = \left(\frac{4\sigma}{c^3} T_r^4\right)\left(\rho_{m0} L^{-3}\right)^{-\frac{4}{3}} \qquad \text{(C-5)}$$

We use the relationship for the adiabatic expansion of radiation: namely,

$$T_r = T_0/L \qquad \text{(C-6)}$$

where T_0 is the present background temperature; then the dependence on L in (C-5) vanishes, and one has

$$\rho_r \rho_m^{-\frac{4}{3}} = \text{constant} \qquad \text{(C-7)}$$

a relationship that holds throughout the expansion.

The spectral distribution of blackbody radiation is given by the Planck function, in frequency units

$$B_\nu d\nu = \frac{2h\nu^3}{c^2} \frac{d\nu}{\exp\left[(h\nu/kT)\right] - 1} \text{ergs}/\left(\text{cm}^2 \text{ sec sr Hz}\right) \qquad \text{(C-8)}$$

where the quantity B_ν is the monochromatic brightness in the frequency interval $d\nu$ at frequency ν, h is the Planck constant, k is the Boltzmann constant, c is the velocity of light, sr (steradian) is the unit solid angle, and Hz (short for Hertz) is unit frequency in cycles per second. We emphasize again that the radiation is characterized by one parameter: the temperature. The maximum brightness in (C-8) occurs at a frequency $\nu_{\max}$, where

$$\frac{Tc}{\nu_{\max}} = 0.510 \text{ cm}^\circ \text{ K} \qquad \text{(C-9)}$$

At long wavelengths λ (low frequency), the radiation brightness in $d\nu$ at ν is given by the Rayleigh-Jeans law: namely

$$B_\nu d\nu \cong \frac{2kt}{\lambda^2} d\nu \text{ ergs}/\left(\text{cm}^2 \text{ sec sr Hz}\right) \qquad \text{(C-10)}$$

At short wavelengths λ (high frequency), the Wien law applies: namely

$$B_\nu d\nu \cong \frac{2hc}{\lambda^3} e^{\frac{-hc}{kT}} d\nu \text{ ergs}/\left(\text{cm}^2 \text{ sec sr Hz}\right) \qquad \text{(C-11)}$$

There is a Wien displacement law which is useful in relating the temperature and the wavelength at maximum brightness: namely

$$T\lambda_{max} = 0.290 \text{ cm}^\circ K \qquad (C-12)$$

Consider λ_{max} as an arbitrary length in the universe whose scale changes with time. Since the temperature of adiabatically expanding blackbody radiation scales as

$$T \propto \frac{1}{L} \qquad (C-13)$$

then clearly

$$L/\lambda = \text{constant} \qquad (C-14)$$

which makes sense of describing the relict blackbody radiation as redshifted. Define redshift as $z = d\lambda/\lambda$. Then

$$1+z = \frac{\lambda+d\lambda}{\lambda} = \frac{L_{now}}{L\left(\text{of time of emission}\right)} \qquad (C-15)$$

Finally, we give some relationships involving the dimensionless entropy per baryon. The heat capacity of blackbody radiation can be written

$$C_r = \frac{16\sigma}{c}T^3 \text{ ergs}\left(\text{cm}^3\text{K}\right)^{-1} \qquad (C-16)$$

while the heat capacity of smeared-out matter present, considered as an ideal gas, is

$$C_m = \frac{3}{2}nk \text{ ergs}\left(\text{cm}^3\text{K}\right)^{-1} \qquad (C-17)$$

where n is the concentration (number density) of baryons. Form the ratio

$$\frac{C_r}{C_m} = 2\left(\frac{16\sigma T^3}{3ckn}\right) \qquad (C-18)$$

and note that T^3 and n both vary as L^{-3}. Hence

$$\frac{C_r}{C_m} = \text{constant during the expansion} \qquad \text{(C-19)}$$

Note also that $16\sigma T^3/(3ck)$ cm^{-3} is the density of dimensionless entropy of the blackbody radiation; that is, the entropy of radiation $16\sigma T^3/(3c)$ divided by the Boltzmann constant k. Hence

$$\frac{C_r}{C_m} = 2\left(\frac{S}{n}\right) \qquad \text{(C-20)}$$

is twice the dimensionless entropy per baryon, or twice the dimensionless entropy in the radiation field per baryon in the universe.

One of the useful attributes of the entropy per baryon is that it is numerically equal, within factors of the order of unity, to the ratio of the pressure of radiation to the pressure of matter as well as the ratio of the concentrations of photons to baryons. Numerically, with current plausible cosmological parameters,

$$\frac{S}{n} \cong 7 \times 10^9 = \text{constant} \qquad \text{(C-21)}$$

Since the value of S/n depends on the values of ρ_m and ρ_r at a given epoch, it is frequently convenient to plot calculated relative abundances in primordial nucleosynthesis against S/n, which in effect contains the one free parameter in the Big Bang model: namely, ρ_m at the onset of nucleosynthesis.

The Age of the Universe

With the present epoch being matter-dominated, one solves the relativistic energy equation to find the time t_0 corresponding to the present epoch (note that the density of radiation now is about 1/1000 the present density of matter in the cosmos). We define a critical total density: that is, the density for the case when the radius of curvature is infinite. We call

$$H_0 = \left(\frac{1}{L}\frac{dL}{dt}\right) \text{now, i.e., } t = t_0, \qquad \text{(A-1)}$$

the current value of the Hubble parameter. Then we obtain

$$\rho_{cr} = \frac{3H_o^2}{8\pi G} \text{ gm cm}^{-3} \qquad \text{(A-2)}$$

It is customary to deal with

$$\Omega = \frac{\rho}{\rho_{cr}} \qquad \text{(A-3)}$$

where it must be remembered that ρ includes

ρ_r (radiation plus relativistic elementary particles)
ρ_m (ordinary baryonic matter plus dark matter)
ρ_s (contribution to Ω due to vacuum energy density; may have driven the expansion during inflation and may, early on, have led to acceleration)

and Ω_0 is then the present value in what follows. The result is for $\Omega_0 > 1$,

$$H_0 t_0 = \frac{-1}{\Omega_0 - 1} + \frac{\Omega_0}{2(\Omega_0 - 1)^{3/2}} \arccos\left(\frac{2 - \Omega_0}{\Omega_0}\right) \qquad \text{(A-4)}$$

for $\Omega_0 = 1$,

$$H_0 t_0 = \frac{2}{3} \qquad \text{(A-5)}$$

for $\Omega_0 < 1$,

$$H_0 t_0 = \frac{1}{1 - \Omega_0} - \frac{\Omega_0}{2(1 - \Omega_0)^{3/2}} \operatorname{arccos h}\left(\frac{2 - \Omega_0}{\Omega_0}\right) \qquad \text{(A-6)}$$

Note that advocates of the inflationary model basically require that $\Omega = 1$ throughout the expansion, where the second of the foregoing relations pertains.

Figure A.1 shows calculations of the present age, presented in units of the Hubble age, as a function of Ω. Note that equations (A-3)–(A-5) apply generally at any t if H and Ω are taken at that t. Also note that the "Hubble age" is given from (A-5) by

$$t_0 = \frac{2}{3H_0} \qquad \text{(A-7)}$$

Figure A.2 shows the present expansion rate plotted against the age of the expansion for an illustrative set of cosmological parameters: namely, present matter and radiation density values. Also shown are the corresponding values of Ω.

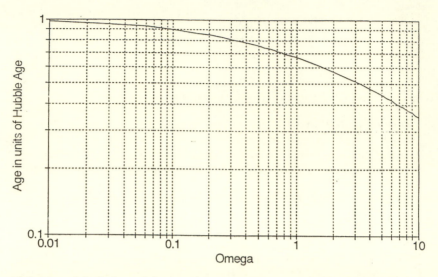

Figure A.1. Calculations of the age of the cosmos in units of the Hubble age $(2/3)$ $(1/H_0)$, where H_0 is the present value of the Hubble parameter, plotted against the density ratio Ω, for $T_{now} = 2.7$ kelvin and with the cosmological constant $\Lambda = 0$.

The Planck Numbers

Given the c.g.s. system of units (centimeters, grams, seconds; or cm, gm, sec; or ℓ, m, t), Planck sought a new system of units ℓ_{PL}, m_{PL}, t_{PL}, based on appropriate constants of nature: namely,

$$c = 3 \times 10^{10} \text{ cm sec}^{-1} \qquad \text{(units } \ell t^{-1})$$
$$G = 6.67 \times 10^{-8} \text{ dyne cm}^2\text{gm}^{-2} \qquad \text{(units } \ell^3 m^{-1} t^{-2}) \qquad \text{(PL-1)}$$
$$\hbar = h/2\pi = 1.055 \times 10^{-27} \text{ erg sec} \qquad \text{(units } \ell^2 m t^{-1})$$

where c is the velocity of light, G is Newton's constant of gravitation, and $\hbar$ is Planck's constant.

One then writes a Planck mass m

$$m_{PL} \text{ (in gm)} = G^x \hbar^y c^z = (\ell^3 m^{-1} t^{-2})^x (\ell^2 m t^{-1})^y (\ell t^{-1})^z \qquad \text{(PL-2)}$$

and solves three simultaneous algebraic equations for x, y, and z:

$$3x + 2y + z = 0$$
$$-x + y = 1 \qquad \text{(PL-3)}$$
$$-2x - y - z = 0$$

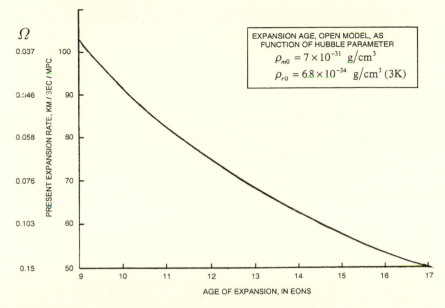

Figure A.2. The expansion age (1 eon = 10^9 years) plotted against the present expansion rate. Corresponding values of Ω are also shown, given the present values of matter and radiation densities, as listed.

giving

$$x = -\tfrac{1}{2},\ y = \tfrac{1}{2},\ z = \tfrac{1}{2}$$

and

$$m_{\mathrm{PL}} = \sqrt{\frac{c\hbar}{G}} = 2.2 \times 10^{-5}\ \mathrm{gm} \tag{PL-4}$$

In the same way, one solves for the other basic Planck units for time and length,

$$t_{\mathrm{PL}} = \sqrt{\frac{G\hbar}{c^5}} = 5.4 \times 10^{-14}\ \mathrm{sec} \tag{PL-5}$$

and

$$l_{\mathrm{PL}} = \sqrt{\frac{G\hbar}{c^3}} = 1.6 \times 10^{-33}\ \mathrm{cm} \tag{PL-6}$$

From these one can write derived quantities as follows: a Planck density,

$$\rho_{PL} \cong m_{PL} l_{PL}^{-3} = 5.4 \times 10^{93} \text{ gm cm}^{-3} \tag{PL-7}$$

a Planck temperature,

$$T_{PL} \text{(temperature)} \cong \frac{m_{PL} c^2}{k} = 1.4 \times 10^{32} \,{}^{\circ}\text{K} \tag{PL-8}$$

a Planck energy,

$$E_{PL} = m_{PL} c^2 = 2.0 \times 10^{16} \text{ ergs} = 1.2 \times 10^{28} \text{ ev} = 1.2 \times 10^{19} \, G\,ev \tag{PL-9}$$

and a Planck power,

$$P_{PL} \text{(power)} \cong \hbar \big/ t_{PL}^2 = \frac{c^5}{G} = 3.6 \times 10^{59} \text{ erg sec}^{-1} \tag{PL-10}$$
$$= 5.6 \times 10^{52} \text{ joules sec}^{-1} = 3.6 \times 10^{52} \text{ watts}$$

In deriving temperature, we have introduced the Boltzmann constant

$$k = 1.38 \times 10^{-16} \text{ ergs(K)}^{-1}$$

In discussing the blackbody radiation and the line element for the early universe, we found that the density of radiation, or relativistic particles (which behave like radiation), is

$$\rho_r = \frac{3}{32 \pi G} \cdot \frac{1}{t^2} = 4.5 \times 10^5 \cdot \frac{1}{t^2} \text{ gm cm}^{-3} \text{ with } t \text{ in seconds} \tag{PL-11}$$

Given the familiar $E = Mc^2$, then the energy density is

$$E_r = \rho_r c^2 \text{ ergs cm}^{-3} \tag{PL-12}$$

Thus the energy content of the universe, taking $R = c \cdot t_{age}$ is

$$E = \frac{4}{3} \pi R^3 \cdot \rho_r c^2 = \frac{c^5}{8G} t \text{ ergs} \tag{PL-13}$$

with t in seconds.

Figure A.3 is an attempt to place the Planck time, which is important in inflationary modeling of the universe, in some kind of schematic context: that is, a time line. Note that the history of the Big Bang is broken up into steps each of which is about 10^{17} times the previous step.

The Uncertainty Principle

We now consider Heisenberg's uncertainty principle, normally applied to quantum mechanical systems but now applied to the early universe; its application to this regime may or may not be appropriate. The principle states that

$$\Delta E \Delta t \geq \hbar \text{ erg sec} \qquad \text{(U-1)}$$

where ΔE is the uncertainty of energy of the system, and Δt is time uncertainty or a characteristic time for the system.

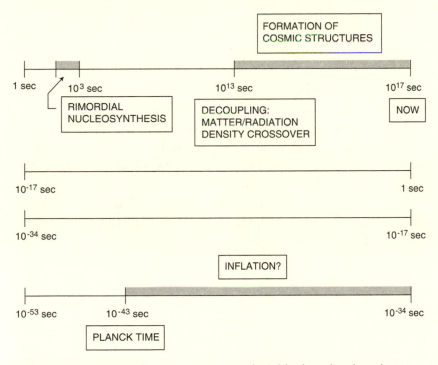

Figure A.3. A time line for the universe on a logarithmic scale, plotted to emphasize the time location of events after one second into the Big Bang with respect to the Planck time.

We recall equation (PL-13), which states the energy content of the universe

$$E = \frac{c^5}{8G} t \text{ ergs} \tag{U-2}$$

Identify this E as ΔE and this t as Δt. Then

$$\Delta E = \frac{c^5}{8G} \Delta t \geq \frac{\hbar}{\Delta t} \text{ ergs} \tag{U-3}$$

Solve for

$$\Delta t = \sqrt{\frac{8G\hbar}{c^5}} \cong 1.5 \times 10^{-43} \text{ sec} \tag{U-4}$$

which is very close to the Planck time (PL-5) and which may or may not be of basic significance.

"Divine Curves": Temporal Behavior of Big Bang Models

Many years ago we found an interesting way to represent the behavior with time of the important characteristics of Big Bang models. Our late colleague George Gamow was intrigued by the graphs we developed, labeled them "divine curves," and used one of them as part of the letterhead of some new personal stationary.

The "divine curves" in figures A.4, A.5, and A.6 are all based on solutions of the Friedmann-Lemaître-Robertson-Walker line element with Einstein's field equations, and share the common parameters

$$H_0 \cong 55 \text{ km sec}^{-1}\text{Mpc}^{-1} \cong 1.78 \times 10^{-18} \text{ sec}^{-1}$$
$$\rho_{r0} \cong 4.47 \times 10^{-34} \text{ gm cm}^{-3} \text{ i.e., } T_0 = 2.725\text{K}$$
$$\rho_{cr} = 5.7 \times 10^{-30} \text{ gm cm}^3$$

Each figure shows the temporal behavior of ρ_r, ρ_m, T, L, and the redshift $(1 + z)$, explicitly shown only for the open model, since $L = \frac{1}{1+z}$.

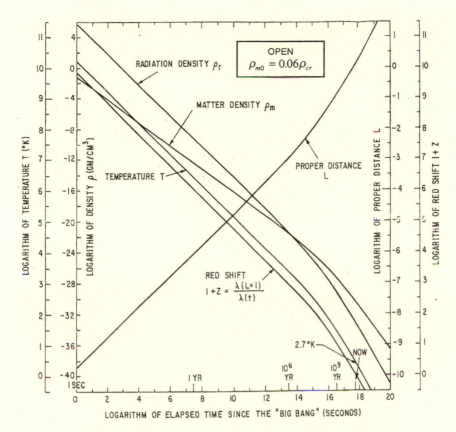

Figure A.4. "Divine curve" (so dubbed by George Gamow) for an *open* cosmological model, with the present matter density equal to 6% of the critical density. This logarithmic plot shows the various cosmological parameters plotted against elapsed time after the Big Bang. Redshift is plotted on this figure only. On figures A.5 and A.6 note that $\frac{1}{H_z} = L$. (Reprinted with permission from *Proceedings of the American Philosophical Society* 119 [1975]: 325–45.)

The scales on the figures are logarithmic, as needed to encompass the enormous range of the parameters.

$\rho_{m0} =$	$0.06\rho_{cr}$	ρ_{cr}	$1.5\rho_{cr}$
$t_0 =$	16.6×10^9 yr.	11.8×10^9 yr.	10.9×10^9 yr.
t_c (when $\rho_r = \rho_m$)	13×10^6 yr.	4,800 yr.	2,140 yr.
T_c	2,000K	34,000K	51,400K
z_c	760	12,700	19,100

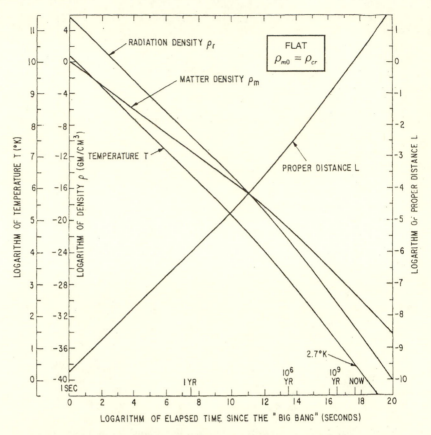

Figure A.5. "Divine curve" for a *flat* cosmological model, with the present matter density equal to the critical density. This logarithmic plot shows the various cosmological parameters plotted against elapsed time after the Big Bang. (Reprinted with permission from *Proceedings of the American Philosophical Society* 119 [1975]: 325–45.)

R_0	$18 \times 18^9 \, i$ light yr.	∞	25×10^9 light yr.
$\rho_m \, (t = 1\text{sec})$	6×10^{-2} gm cm^{-3}	0.63 gm cm^{-3}	1.5 gm cm^3
$\rho_r \, (t = 1\text{sec})$	4.49×10^5 gm cm^{-3}	4.49×10^5 gm cm^{-3}	4.49×10^5 gm cm^{-3}
$\left(\dfrac{dT}{dt}\right)_0$	-4.8×10^{-18}K sec^{-1}	-4.8×10^{-18}K sec^{-1}	-4.8×10^{-18}K sec^{-1}

Finally, these three figures tend to emphasize visually what happened early on. To show the behavior at longer times, as well as during one cycle for the "closed" case, refer to figure 6.1, showing the scale factor plotted against the elapsed time since the Big Bang on a time scale very different from that of figures A.4–A.6.

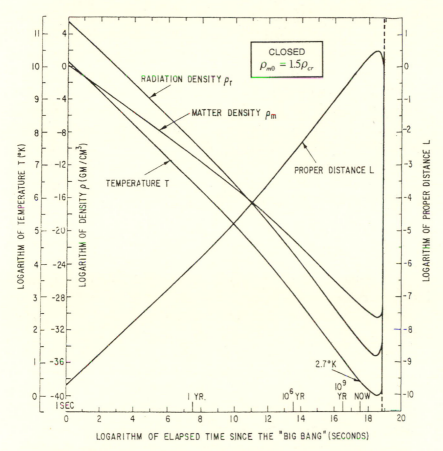

Figure A.6. "Divine curve" for a *closed* cosmological model, with the present matter density equal to 1.5 times the critical density. This logarithmic plot shows the various cosmological parameters plotted against elapsed time after the Big Bang. (Reprinted with permission from *Proceedings of the American Philosophical Society* 119 [1975]: 325–45.)

The Density Ratio Omega in the Standard Big Bang Model

The density ratio Ω plays an important role in cosmology. If the canonical Big Bang model is valid, then ultimately the value of Ω will tell us whether the universe is open, closed, or exactly poised between. In a homogeneous, isotropic universe with a smoothed-out density, Ω is defined as

$$\Omega = \rho/\rho_{cr} \qquad (\Omega\text{-}1)$$

where ρ is the mean mass density of matter and radiation in the universe; ρ_{cr} is the critical density, the density for closure (or, if you will, the density that pertains when the radius of curvature of the universe is infinite); and both densities are functions of the epoch. Usually in dealing with the standard Big Bang model, one assumes no matter-radiation conversion, but one can show that interconversion does not affect the behavior of H and Ω in the early universe. We also note that $\Omega = 2q_0$ where q_0 is the usual deceleration parameter (which we do not discuss here). The critical density divides the solution space of the Friedmann-Lemaître equation into an open and a closed model, and, with the cosmological constant taken as zero here and in what follows, is defined as

$$\rho_{cr} = H^2/\gamma \qquad (\Omega\text{-}2)$$

where H is the Hubble parameter (epoch-dependent), γ is $8\pi G/3$, and G is the constant of gravitation.

We approach the development of early-time approximations for Ω by rearranging the usual Friedmann-Lemaître equation to read

$$\rho_t = \gamma^{-1}[H^2 - c^2/(L^2 R_i^2)] \qquad (\Omega\text{-}3)$$

where we have replaced the radius of curvature R_0 by iR_i, thereby selecting the open cosmological model for further consideration. The quantity $L = \ell/\ell_0$ is a dimensionless proper distance or scale factor for the expansion, where ℓ_0 is chosen so that $L = 1$ at the present epoch. Combining equations (Ω-2) and (Ω-3) then yields

$$\Omega = 1 - c^2/(R_i^2 L^2 H^2) \qquad (\Omega\text{-}4)$$

To evaluate the second term of (Ω-4), we use the exact parametric solutions of the Friedmann-Lemaître equation, again selecting for an open model with no approximation, which can be written, with ϕ as a "development angle"

$$L = a_0(\cosh\phi - 1) + b_0 \sinh\phi \text{ cm} \qquad (\Omega\text{-}5)$$

and

$$t = (R_i/c)[a_0(\sinh\phi - \phi) + b_0(\cosh\phi - 1)] \text{ sec} \qquad (\Omega\text{-}6)$$

where the dimensionless coefficients a_0 and b_0 are

$$a_0 = (c^2 R_i^2 \kappa \rho_{m0})/6 \text{ and } b_0 = cR_i\gamma(\rho_{r0})^{1/2} \qquad (\Omega\text{-}7)$$

with $\kappa = 8\pi G/c^4$ and ρ_{m0} and ρ_{r0} the present mass densities of matter and radiation, respectively.

We now differentiate these solutions with respect to the development angle ϕ, with the result

$$dL = (a_0 \sinh\phi + b_0)d\phi \qquad (\Omega\text{-}8)$$

and

$$dt = (R_i/c)[a_0(\sinh\phi - \phi) + b_0(\cosh\phi - 1)]d\phi \qquad (\Omega\text{-}9)$$

This enables us to write

$$\Omega = 1 - \{[(A \cosh\phi - 1) + \sinh\phi]^2/[A \sinh\phi + \cosh\phi]^2\} \qquad (\Omega\text{-}10)$$

where

$$A = a_0/b_0 \qquad (\Omega\text{-}11)$$

Alternatively, one can replace L^2H^2 in (Ω-4) with the result

$$\Omega = 1 - \{\gamma(R_i^2/c^2)[(\rho_{m0}/L) + \rho_{r0}/L^2] + 1\}^{-1} \qquad (\Omega\text{-}12)$$

which can be evaluated directly, again using the exact parametric solutions for L and t. We note that equations (Ω-10) and (Ω-12) are both valid for all epochs in the expansion and can be shown to be completely equivalent.

Early-Time Approximation. Equation (Ω-12) is particularly convenient for examining the behavior of Ω at early times in the expansion, although we emphasize again that (Ω-10) can also be used as a basis. At sufficiently early time, we could in principle discard the term involving $1/L$, which requires that $\rho_{r0}/L^2 >> \rho_{m0}/L$. This is equivalent to confining one's attention to times prior to the time of matter-radiation decoupling, where $L = L_c$, with $L_c = \rho_{m0}/\rho_{r0}$ so that $L(\text{early}) = L_e << L_c$. One can integrate the Friedmann-Lemaître equation at early time by neglecting the term containing the radius of curvature compared with the term containing the radiation density. First we obtain an approximation for $L(early) = L_e$ as a function of time: namely,

$$L_e^2 \approx (4\gamma\rho_{r0})^{1/2}t \qquad (\Omega\text{-}13)$$

or, for purposes of approximation in (Ω-12),

$$\rho_{r0}/L_e^2 \approx [\rho_{r0}/4\gamma)]^{1/2}t^{-1} \qquad (\Omega\text{-}14)$$

Then, the early-time approximation for Ω follows, using (Ω-13) in (Ω-12), with the result

$$\Omega \approx 1 - (2c^2/R_i^2)(\gamma\rho_{r0})^{-1/2}t \approx 1 - \{[R_i^2/(2c^2)](\gamma\rho_{r0})^{1/2} + t\}^{-1}\,t \qquad (\Omega\text{-}15)$$

In (Ω-15) we ask, for what value of t can we neglect the second term (i.e., t alone) with respect to the first term in the bracket in the denominator? For $H_0 = 50$ km/(secMpc), the first term has the numerical value 3.22×10^{15} sec, so that $t(early) \ll 3.22 \times 10^{15}$ sec and

$$\Omega_{e,50} \approx 1 - 3.1 \times 10^{-16}t \qquad (\Omega\text{-}16)$$

while for $H_0 = 75$ km/(secMpc), $t(early) \ll 1.39 \times 10^{15}$ sec, so that

$$\Omega_{e,75} \approx 1 - 7.2 \times 10^{-16}t \qquad (\Omega\text{-}17)$$

It is important to note that these early-time approximations are indistinguishable from the exact expression for Ω in the time range in which they are valid.

It appears that the value of Ω in the standard model is extremely close to unity with no special assumptions being required. If there is any validity to the concept of a Planck time, 10^{-43} seconds, and if the physics of the Big Bang model is still valid then, the deviation from unity is about one part in 10^{60} and is still one part in 10^{46} at 10^{-30} seconds, usually identified as the end of the inflationary era. Thus one has the extremely small deviation required by various inflationary paradigms, which also purport to involve conventional physics back to the Planck time. But it appears to be provided by the standard Big Bang model, as well.

In addition, it is a part of the inflationary paradigm that Ω be unity throughout the expansion. In fact, the exact solution for Ω in the standard Big Bang model is unity from the Planck time out to times of the order of a million years after the start of the expansion, and then it deviates, finally, to encompass the value preferred for Big Bang nucleosynthesis.

In sum, although it may be that one has to have inflation early in order to deal with such problems as finding an isotropic distribution of the cosmic microwave background radiation—including regions that could not have been causally connected by the time of crossover—as well as the apparent nonexistence of monopoles, it does not seem necessary to discard the standard Big Bang model during the early epochs in order to have $\Omega = 1$ with high precision.

Recommended Reading

This list (compiled December 1999) is admittedly incomplete, being limited to items with which we are familiar. No value judgment is implied for those books not included.

Treatises with Considerable Technical Content

Barrow, J. D., and F. J. Tipler. *The Anthropic Cosmological Principle*. New York: Oxford University Press, 1986.

Bernstein, J., and G. Feinberg. *Cosmological Constants: Papers in Modern Cosmology*. New York: Columbia University Press, 1986.

Bertotti, B., R. Balbinot, S. Bergia, and A. Messina, eds. *Modern Cosmology in Retrospect*. New York: Cambridge University Press, 1990.

Coles, P., and F. Lucchin. *Cosmology: The Origin and Evolution of Cosmic Structure*. New York: John Wiley, 1995.

Einstein, A. *The Meaning of Relativity*. Princeton, NJ: Princeton University Press, 1946.

Harwit, M. *Astrophysical Concepts*. 3d edition. New York: Springer-Verlag, 1988.

Hawley, J. F., and K. A. Holcomb. *Foundations of Modern Cosmology*. New York: Oxford University Press, 1998.

Hetherington, N. S., ed. *Encyclopedia of Cosmology*. New York: Garland, 1993.

Kolb, E., and M. Turner. *The Early Universe*. Reading, MA: Addison-Wesley, 1990.

Kragh, H. *Cosmology and Controversy: The Historical Development of Two Theories of the Universe*. Princeton, NJ: Princeton University Press, 1996.

Misner, C. W., K. S. Thorne, and J. A. Wheeler. *Gravitation*. San Francisco: W.H. Freeman, 1973.

Ohanian, H. C. *Gravitation and Spacetime*. New York: W.W. Norton, 1976.

Partridge, B. *3K: The Cosmic Microwave Background Radiation*. New York: Cambridge University Press, 1992.

Peebles, P. J. E. *Principles of Physical Cosmology*. Princeton, NJ: Princeton University Press, 1993.

Rees, M. T. *Perspectives in Astrophysical Cosmology*. Cambridge: Cambridge University Press, 1995.

Sandage, A. R., R. G. Kron, and M. S. Longair, contributors. *The Deep Universe*, ed. B. Binggeli and R. Buser. Berlin: Springer-Verlag, 1995.

Schramm, D. N. *The Big Bang and Other Explosions in Nuclear and Particle Astrophysics*. Singapore: World Scientific, 1996.

Tolman, R. C. *Relativity, Thermodynamics, and Cosmology*. Oxford: Clarendon Press, 1934.

Weinberg, S. *Gravitation and Cosmology: Principles and Applications of the General Theory of Relativity*. New York: John Wiley, 1972.

Technical Papers

Adams, F. C., and G. Laughlin. "A Dying Universe: The Long-Term Fate and Evolution of Astrophysical Objects." *Reviews of Modern Physics* 69 (1997): 337. Also see popular-level article by these authors in *Sky and Telescope* (August 1998).

Alpher, R. A., and G. Marx. "The Creation of Free Energy." *Vistas in Astronomy* 35 (1992): 179.

Dyson, F. J. "Time without End: Physics and Biology in an Open Universe." *Reviews of Modern Physics* 51 (1979): 447.

Books for a More General Audience

Alfvén, H. *Worlds-Antiworlds: Antimatter in Cosmology*. San Francisco: W.H. Freeman, 1966.

Barrow, J. D., and J. Silk. *The Left Hand of Creation: The Origin and Evolution of the Expanding Universe*. New York: Basic Books, 1983.

Bartusiak, M. *Thursday's Universe*. New York: Times Books, 1986.

Bartusiak, M. *Through a Universe—Darkly*. New York: HarperCollins, 1993.

Brockelman, P. *Cosmology and Creation: The Spiritual Significance of Contemporary Cosmology*. New York: Oxford University Press, 1999.

Chaisson, E. *Cosmic Dawn: The Origins of Matter and Life*. New York: W.W. Norton, 1981.

Charlesworth, K., and J. Gribben. *The Cartoon History of Time*. New York: Penguin Books, 1990.

Chown, M. *Afterglow of Creation*. Sausalito, CA: University Science Books, 1996.

Coles, P., and G. F. R. Ellis. *Is the Universe Open or Closed? The Density of Matter in the Universe.* New York: Cambridge University Press, 1997.

Dauber, P. M., and R. A. Muller. *The Three Big Bangs: Comet Crashes, Exploding Stars, and the Creation of the Universe.* Reading, MA: Addison-Wesley, 1996.

Delsemme, A. H. *Our Cosmic Origins: From the Big Bang to the Emergence of Life and Intelligence.* Cambridge: Cambridge University Press, 1998.

Ferris, T. *The Whole Shebang: A State-of-the-Universe Report.* New York: Simon and Schuster, 1997.

Gamow, G. *One Two Three . . . Infinity.* New York: Viking Press, 1947.

Gamow, G. *The Creation of the Universe.* New York: Viking Press, 1952.

Gleiser, M. *The Dancing Universe: From Creation Myth to the Big Bang.* New York: Dutton, Penguin, 1997.

Goldsmith, D. *Einstein's Greatest Blunder: The Cosmological Constant and Other Fudge Factors in the Physics of the Universe.* Cambridge, MA: Harvard University Press, 1995.

Greene, B. *The Elegant Universe.* New York: W.W. Norton, 1999.

Gribben, J. *In Search of the Big Bang.* New York: Bantam Books, 1986.

Guth, A. *The Inflationary Universe.* Reading, MA: Addison-Wesley, 1997.

Harrison, E. R. *Cosmology: The Science of the Universe.* New York: Cambridge University Press, 1986.

Harrison, E. R. *Masks of the Universe.* London: Collier, Macmillan, 1985.

Hawking, S. *A Brief History of Time: From the Big Bang to Black Holes.* New York: Bantam Books, 1988.

Hawking, S. *The Illustrated Brief History of Time.* New York: Bantam Books, 1996.

Hogan, C. J. *The Little Book of the Big Bang: A Cosmic Primer.* New York: Springer-Verlag, 1998.

Hoskin, M., ed. *The Cambridge Illustrated History of Astonomy.* Cambridge: Cambridge University Press, 1997.

Islam, J. N. *The Ultimate Fate of the Universe.* New York: Cambridge University Press, 1983.

Kaku, M., and J. Trainer. *Beyond Einstein: The Cosmic Quest for the Theory of the Universe.* New York: Bantam Books, 1987.

Koestler, A. *The Sleepwalkers.* New York: Grosset and Dunlap, 1963.

Kolb, E. W. *Blind Watchers of the Sky: The People and Ideas That Shaped Our View of the Universe.* Reading, MA: Addison-Wesley, 1996.

Lankford, J., ed. *History of Astronomy: An Encyclopedia.* New York: Garland, 1997.

Lemaître, G. *The Primeval Atom: A Hypothesis of the Origin of the Universe.* Trans. B. H. Korff and S. A. Korff. New York: D. Van Nostrand, 1950.

Lightman, A., and R. Brawer. *Origins: The Lives and Worlds of Modern Cosmologists.* Cambridge, MA: Harvard University Press, 1990.

Longair, M. S. *Our Evolving Universe*. Cambridge: Cambridge University Press, 1996.

Mather, J. C., and J. Boslough. *The Very First Light*. New York: Basic Books/HarperCollins, 1996.

Merleau-Ponty, J., and B. Morando. *The Rebirth of Cosmology*. New York: Alfred A. Knopf, 1976.

Morris, R. *The End of the World*. New York: Anchor Press/Doubleday, 1980.

Munitz, M. K. *Space, Time and Creation: Philosophical Aspects of Scientific Cosmology*. Glencoe, IL: Free Press, 1957.

Munitz, M. K. *Theories of the Universe*. Glencoe, IL: Free Press, 1957.

Narlikar, J. V. *Seven Wonders of the Cosmos*. New York: Cambridge University Press, 1999.

Overbye, D. *Lonely Hearts of the Cosmos: The Scientific Quest for the Secret of the Universe*. New York: HarperCollins, 1991.

Padmanabhan, T. *After the First Three Minutes: The Story of Our Universe*. New York: Cambridge University Press, 1998.

Rees, M. J. *Before the Beginning: Our Universe and Others*. Reading, MA: Addison-Wesley Longman, 1997.

Riordan, M., and D. M. Schramm. *The Shadows of Creation: Dark Matter and the Structure of the Universe*. New York: W.H. Freeman, 1991.

Rowan-Robinson, M. *Ripples in the Cosmos: A View Behind the Scenes of the New Cosmology*. New York: W.H. Freeman/Spektrum, 1993.

Silk, J. *The Big Bang*. 2d edition. San Francisco: W.H. Freeman, 1995.

Smith, R. W. *The Expanding Universe: Astronomy's "Great Debate," 1900–1931*. New York: Cambridge University Press, 1982.

Smolin, L. *The Life of the Cosmos*. New York: Oxford University Press, 1997.

Smoot, G., and K. Davidson. *Wrinkles in Time*. New York: Wm. Morrow, 1993.

Weinberg, S. *The First Three Minutes*. Rev. ed. New York: Basic Books/ HarperCollins, 1988.

Index

chemical elements. *See* elements
closed universe model
　critical density in, 110, 160–162, 193–194
　entropy trends in, 110, 157
　inflationary era and, 136–137
　radiation predictions for, 163
clusters. *See also* galactic clusters; star
　　clusters
　as cosmos building blocks, 9–10
CN radical, 115–118
COBE (Cosmic Background Explorer)
　　satellite
　background radiation and, 108, 112–113,
　　123; measurements of, 101–102, 123–
　　126, 181
　in Big Bang evidence, 25, 32, 82, 91
　inflationary era radiation and, 133–134
　launching of, 122–123
cold matter, 93
comoving elements, 96
computers, theory development with, 71–
　72, 76, 79, 85
conservation of charge principle, 4
constants of nature
　in anthropic principle, 104–105, 169–170
　as Big Bang alternative, 102–104
　Planck numbers and, 142–143
　properties of, 10–11, 96, 159, 169
　vs. universe, as first, 140
Continuous Creation theory, 20, 95–99
contraterrene matter, 23
coordinate points, relativity theory and, 7–
　8, 52
Copernicus, Nicholas, 42–43, 168
cosmic microwave background radiation
　　(CMBR)
　acoustic waves and, 114, 124, 127–128
　alternatives for, 98, 101–102
　anisotropy of, 114, 124–128
　in Big Bang theory, 33, 37, 61–62, 93,
　　163
　as blackbody radiation, 109, 123, 181,
　　188
　COBE satellite and, 101–102, 108, 112–
　　113, 123–126, 181
　discovery of, 9, 13, 33, 45, 107–108;
　　delayed, 114–115, 118–119; impact of,
　　109–115, 118
　equations for, 177, 181–184
　inflationary era and, 133–134
　measurements of: with COBE, 101–102,
　　123–126, 181; early, 115–127; recent,
　　118, 127–129, 164
　origin of, 45, 112–115

in Steady State theory, 98, 152
　structure formation and, 90–91
　Sunyaev-Zeldovich effect of, 58, 62, 127,
　　129
　temperature perspectives of, 65, 120–122
cosmic rays
　origination of, 66–67, 154
　radiation in, 23, 129
cosmic structures. *See also specific*
　　structure
　development theories of, 70, 89–93, 110,
　　126
　seeds of, 42, 91, 125, 134, 145
cosmological constant, 50, 52, 60, 138–139,
　　176
　in static *vs.* nonstatic model, 65–66
cosmological distance ladder, 57
cosmological principle, 8–9, 11–12, 18, 51,
　　96
cosmology, physical
　alternatives to, 95–106, 147–158
　before Big Bang model, 39–50
　Dirac's law of large numbers for, 102–
　　103
　geometric observations of, 7–9, 11–12
　perfect principle for, 8–9, 51, 96
　relativity theories and, 5–8, 11, 47–51
　technological advances in, 40–41, 134
　before 20th century, 41–47
　in 20th century, 4, 40–41, 47–50, 107,
　　173
　visible universe models of, 4–7, 144
coupling constants, 143
creation field (c-field), 96, 151
critical density (Ω)
　in Big Bang model, 85–88, 193–197
　in closed *vs.* open universe, 110, 160–
　　162, 193–194
　definition of, 160, 193–194
　in energy equations, 179, 187
　inflationary era and, 64, 136–139
　intelligent life and, 104
　of radiation, 184–185
　unity trends of, 136–138
　in universe age, 62–64, 160, 184–186
Curtis, Heber D., 44, 59
curvature of space. *See* space-time
cyanogen. *See* CN radical

dark matter
　indirect measurement of, 17, 19
　in Steady State theory, 152–153
　as universe component, 4, 9, 86, 164
Darwin, Charles, 13, 41